Arthur Cockshott, F. B. Walters

A Treatise on Geometrical Conics

In Accordance with the Syllabus of the Association for the Improvement of

Geometrical Teaching

Arthur Cockshott, F. B. Walters

A Treatise on Geometrical Conics
In Accordance with the Syllabus of the Association for the Improvement of Geometrical Teaching

ISBN/EAN: 9783337168001

Printed in Europe, USA, Canada, Australia, Japan

Cover: Foto ©Andreas Hilbeck / pixelio.de

More available books at **www.hansebooks.com**

ON

GEOMETRICAL CONICS

IN ACCORDANCE WITH THE SYLLABUS
OF THE ASSOCIATION FOR THE IMPROVEMENT
OF GEOMETRICAL TEACHING.

BY

ARTHUR COCKSHOTT, M.A.,
ASSISTANT MASTER AT ETON COLLEGE, FORMERLY FELLOW AND ASSISTANT
TUTOR OF TRINITY COLLEGE, CAMBRIDGE,

AND

REV. F. B. WALTERS, M.A.,
PRINCIPAL OF KING WILLIAM'S COLLEGE, ISLE OF MAN, AND FELLOW OF
QUEENS' COLLEGE, CAMBRIDGE.

London:
MACMILLAN AND CO.
AND NEW YORK.
1889

[*All Rights reserved.*]

PREFACE.

THE need of some recognized sequence of propositions in Elementary Geometrical Conics has long been very generally admitted. This need the Association for the Improvement of Geometrical Teaching has attempted to supply by the publication of the Syllabus of Geometrical Conics, which was drawn up by an influential Committee and accepted by the Association at their annual General Meeting in January, 1884.

In the following pages we have given proofs of the propositions in the hope that they may be found useful to those teachers who desire to adopt the order to which the Association has given the weight of its approval.

We have introduced a chapter on Orthogonal Projection immediately after that on the Parabola, as we think it important that the student should understand as early as possible the close connection between the ellipse and circle and should be introduced at once to a method by which so many properties of the ellipse may be deduced from well-known properties of the circle.

At the end of the book will be found a large collection of Cambridge problems; we have given a list of important properties of Conics, not included in the propositions in the text—all of which are considered as well known and may therefore be assumed in the solution of any other problems.

A. C.
F. B. W.

May, 1889.

TABLE OF CONTENTS.

	PAGE
PARABOLA	1
ORTHOGONAL PROJECTIONS	26
ELLIPSE	33
HYPERBOLA	74
RECTANGULAR HYPERBOLA	120
SECTIONS OF A CYLINDER AND CONE	121
ADDITIONAL PROPOSITIONS	144
PROBLEMS	148

PARABOLA.

DEF. I. A *parabola* is the locus of a point (P), whose distance from a fixed point (S) is equal to its distance (PM) from a fixed straight line (XM),
$$(SP = PM).$$

II. The fixed point (S) is called the *focus*.

III. The fixed straight line (XM) is called the *directrix*.

DEF. A curve is *symmetrical with respect to a straight line*, if, corresponding to any point on the curve, there is another point on the curve on the other side of the straight line such that the chord joining them is bisected at right angles by the straight line.

DEF. The straight line is called an *axis* of the curve.

DEF. A *vertex* is a point at which an axis meets the curve.

Proposition I.

Construction for points on the parabola. The perpendicular on the directrix through the focus is an axis of symmetry.

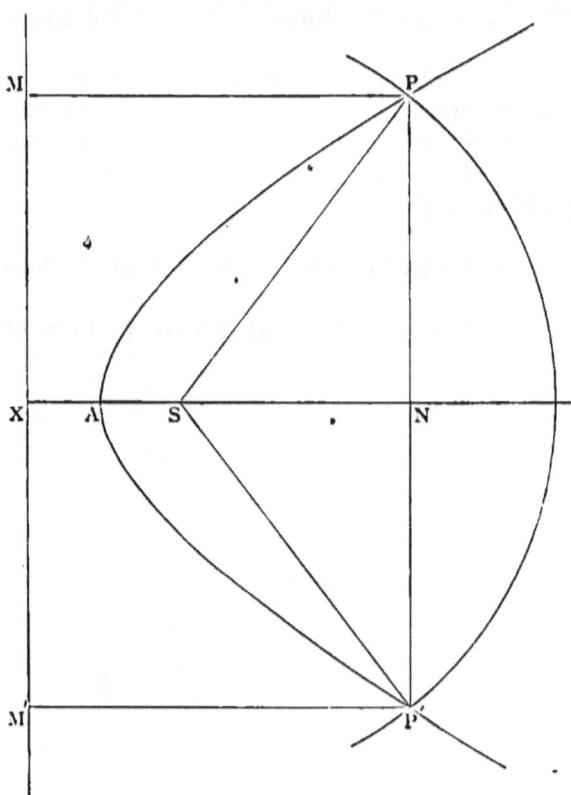

Let S be the focus and MXM' the directrix. Through S draw a straight line SX perpendicular to the directrix, and produce it indefinitely in the direction XS.

Bisect SX in A; then because $SA = AX$, A is a point on the parabola.

In XS or XS produced take any point N; through N draw a straight line PNP' perpendicular to XN; with centre S and radius equal to XN describe a circle, to cut (if possible) PNP' at P and P'; and draw PM, $P'M'$ perpendicular to the directrix.

Then because $$SP = NX = PM,$$
therefore P is a point on the parabola.

Similarly P' is a point on the parabola.

Since $$NP = NP', \qquad \text{[Euc. III. 3.}$$
PP' is bisected at right angles by XS, and the curve is symmetrical with respect to XS.

(1) If N and S lie on the same side of A, SN is less than NX, and the circle will cut the line PNP'.

(2) If N and S lie on opposite sides of A, the circle will not cut the straight line PNP'.

Hence the parabola is unlimited in extent, but lies entirely on one side of a line through A perpendicular to AS.

For riders see p. 7.

DEF. The *axis* (SX) *of a parabola* is a straight line through the focus perpendicular to the directrix.

DEF. The *vertex* (A) *of a parabola* is the point at which the axis meets the curve.

DEF. The *ordinate* (PN) of a point on a parabola is the perpendicular from the point (P) upon the axis.

DEF. The *abscissa* (AN) is the portion of the axis between the vertex and the ordinate.

DEF. The *focal distance* (SP) of a point on a parabola is its distance from the focus.

Proposition II.

If the chord PP′ *intersects the directrix in* K, SK *bisects the exterior angle between* SP *and* SP′.

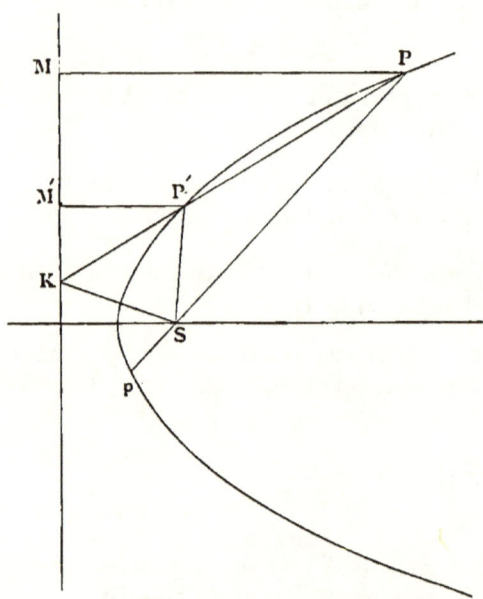

Join SP, SP'.

Draw PM, $P'M'$ perpendicular to the directrix, and produce PS to p.

Then, by similar triangles PKM, $P'KM'$,
$$PK : P'K = PM : P'M'$$
$$= SP : SP';$$

∴ SK bisects the exterior angle $P'Sp$. [Euc. VI. A.

Proposition III.

If PN *is an ordinate to the parabola at the point* P, *then*
$$PN^2 = 4AS \cdot AN.$$

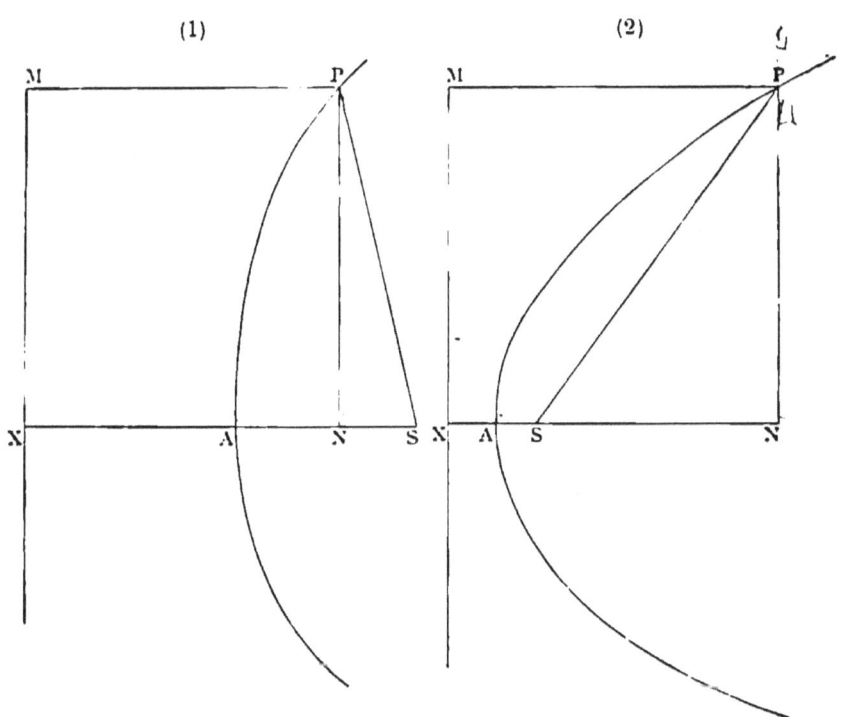

(1) (2)

Join SP, and draw PM perpendicular to the directrix.

Then, since $\qquad SA = AX$,

and AS is divided in N (Fig. 1), AN is divided in S (Fig. 2);

$\qquad\qquad \therefore NX^2 = SN^2 + 4AS \cdot AN.$ [Euc. II. 8.

But $\qquad\qquad NX^2 = PM^2$
$\qquad\qquad\qquad = SP^2$
$\qquad\qquad\qquad = PN^2 + SN^2;$
$\therefore PN^2 + SN^2 = SN^2 + 4AS \cdot AN;$
$\qquad \therefore PN^2 = 4AS \cdot AN.$

DEF. The double ordinate through the focus is called the *latus rectum* (*LL'*).

PROPOSITION IV.

The latus rectum LL' = 4AS.

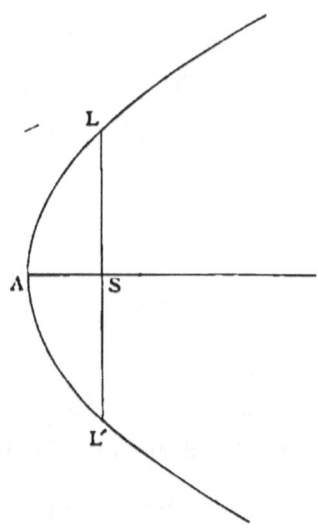

$SL^2 = 4AS \cdot AS$ [Prop. 3.

∴ $SL = 2AS$;

∴ $LL' = 4AS$.

PROBLEMS.

Prop. I.

1. To trace the parabola by points by means of Euc. I. 23.

2. PP', QQ' are double ordinates to the parabola. Shew that PQ, $P'Q'$ meet the axis in the same point.

3. If SM meets the parallel through A to the directrix in Y, shew that SM is bisected in Y.

4. Shew also that PY is perpendicular to SM and bisects angle SPM.

5. SZ is drawn perpendicular to SP to meet directrix in Z. Shew that PZ bisects the angle SPM.

6. If two focal chords of a parabola are equal, they are equally inclined to the axis.

7. Find locus of centre of a circle which touches a given straight line, and passes through a given point.

8. Find locus of centre of a circle which touches a given circle and a given straight line.

9. A straight line parallel to the axis meets the parabola in one point only.

Prop. II.

1. Pp is a focal chord of a parabola, Q any other point on the curve. If PQ, pQ meet the directrix in K and K' respectively, KSK' is a right angle.

2. PQ, pq are focal chords. Shew that Pp, Qq, meet on the directrix. As also do Pq, pQ.

3. If they meet the directrix in K and K', KSK' is a right angle.

4. Trace the parabola by means of this proposition, by joining A to different points in the directrix.

5. P is any point on the parabola. If PA produced meet the directrix in K, MSK is a right angle.

6. Given a parabola and its focus, find the directrix.

7. PQ is a double ordinate of the parabola, PX cuts the curve in P': prove that $P'Q$ passes through the focus.

Prop. III.

1. PP' is a double ordinate of the parabola. If the circle round PAP' cut the axis again in Q, shew that NQ is constant and find its length.

2. PNP' is a double ordinate of the parabola. Through Q, another point on the parabola, straight lines are drawn, one passing through the vertex, and the other parallel to the axis, cutting PP' in L and L'. Shew that $NL \cdot NL' = PN^2$.

Prop. IV.

1. Find a double ordinate PP' of a parabola which shall be double the latus rectum.

2. The radius of the circle described about the triangle $LAL' = \frac{5}{3}$ latus rectum.

DEF. Let PP' be the chord of any curve. Then if the point P' move up to P, the chord PP' in the limiting position when P' coincides with P is called the *tangent* at P.

PROPOSITION V.

If the tangent at P meets the directrix in Z, PSZ is a right angle, and the tangent at P bisects the angle between the focal distance SP and the perpendicular PM on the directrix; and the tangent at the vertex is at right angles to the axis.

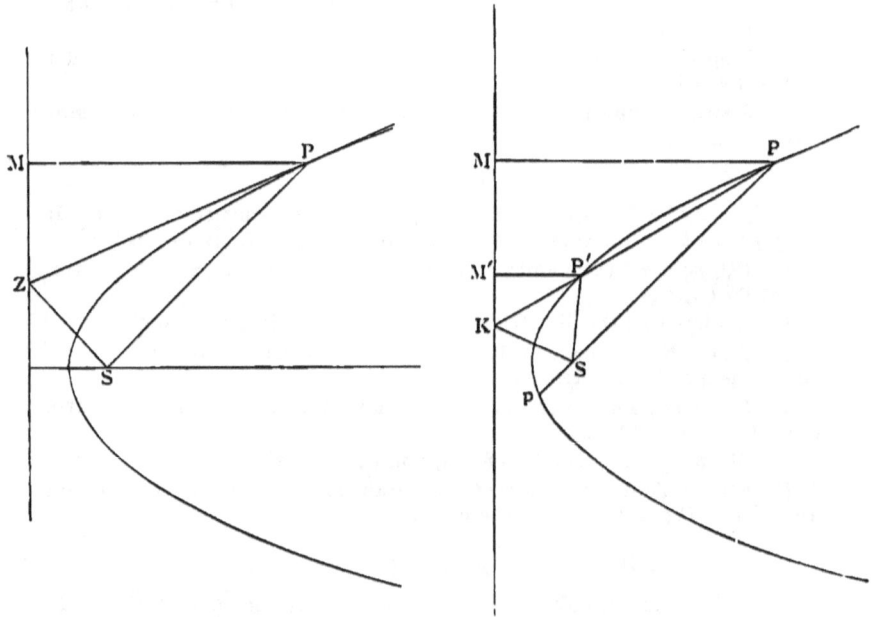

In the figure of Prop. II. let the chord $PP'K$ become the tangent PZ by moving the point P' up to P, then ultimately SK coincides with SZ, SP' coincides with SP, and the angle $P'Sp$ becomes two right angles; but $P'SK$ is always half the angle $P'Sp$ (Prop. II.), hence PSZ is half of two right angles, or PSZ is a right angle.

PARABOLA.

Draw PM perpendicular to the directrix,

$$PM^2 + MZ^2 = PZ^2 \qquad \text{[Euc. I. 47.}$$
$$= SP^2 + SZ^2;$$
$$\therefore MZ^2 = SZ^2, \text{ since } PM = SP;$$
$$\therefore MZ = SZ;$$

∴ in the triangles ZPM, ZPS,

PM, $MZ = PS$, SZ each to each,

and PZ is common to both;

$$\therefore \text{ the angle } MPZ = SPZ. \qquad \text{[Euc. I. 8.}$$

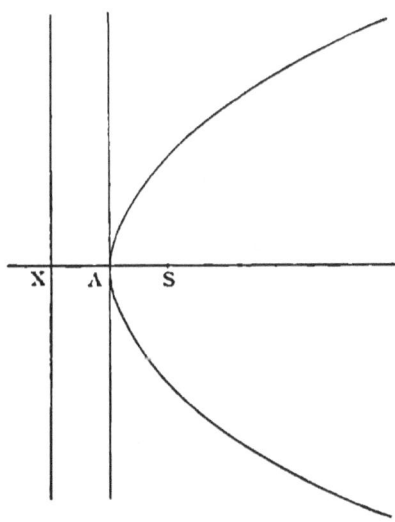

If the point P be at the vertex A, the angle SPM is two right angles and coincides with the straight angle SAX. Hence the tangent, which bisects this angle, is at right angles to the axis.

Prove from the definition of a parabola that the straight line which bisects the angle SPM cannot meet the curve in a second point.

For riders see p. 11.

Proposition VI.

The tangents at the extremities of a focal chord intersect at right angles on the directrix.

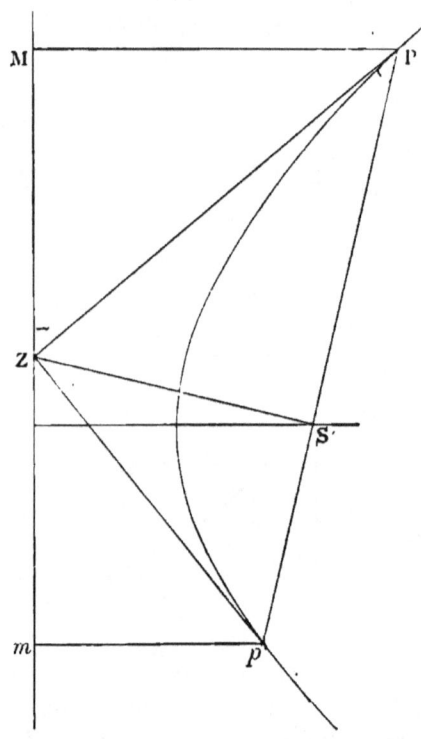

Let PSp be a focal chord, and let the tangent at P intersect the directrix in Z.

Join ZS, Zp, and draw PM, pm perpendicular to the directrix.

PARABOLA. 11

Then, ∵ PZ is the tangent at P,
∴ SZ is at right angles to PSp; [Prop. 5.
∴ pZ is the tangent at p.

Again, ∵ the $\triangle SPZ = \triangle MPZ$, [Euc. I. 4.
∴ $\angle SZP = \angle PZM$;
∴ SZP is half of SZM.

Similarly SZp is half of SZm,
∴ PZp is half of SZM and SZm together,
is half of two right angles;
∴ PZp is a right angle.

PROBLEMS.

Prop. V.

1. The tangents at the extremities of the latus rectum meet the directrix at the point X.

2. If any point O be taken on the tangent at P, $OM = OS$.

3. If the tangents to the parabola at P and P' meet in O, and PM, $P'M'$ be the perpendiculars on the directrix from P and P', OM, OS, OM' are all equal.

Deduce a construction for drawing the two tangents from an external point O.

4. If two tangents OQ, OQ' be drawn to a parabola, and V be the middle point of QQ', prove that OV is parallel to the axis.

5. Hence, given two tangents to a parabola, and their points of contact, to find the focus.

6. If the tangent at P meet the latus rectum produced in K, and the directrix in Z, $SK = SZ$.

Prop. VI.

1. If the tangents at the extremities of the focal chord PP_1 meet in Z and PM, P_1M_1 be perpendiculars on directrix, shew that MM_1 is bisected in Z. Hence, prove that the circle described on PP_1 as diameter touches directrix in Z.

2. PSQ is a focal chord. QG perpendicular to the tangent at Q cutting axis in G. GZ is a perpendicular on the tangent at P. Shew that Z lies on the latus rectum.

3. Tangents at the extremities of a focal chord cut off equal intercepts on the latus rectum.

PARABOLA.

DEF. The straight line which is drawn through any point on a curve at right angles to the tangent at that point is called the *normal*.

PROPOSITION VII.

If the tangent and normal at P *meet the axis at* T *and* G *respectively,* $SG = SP = ST$.

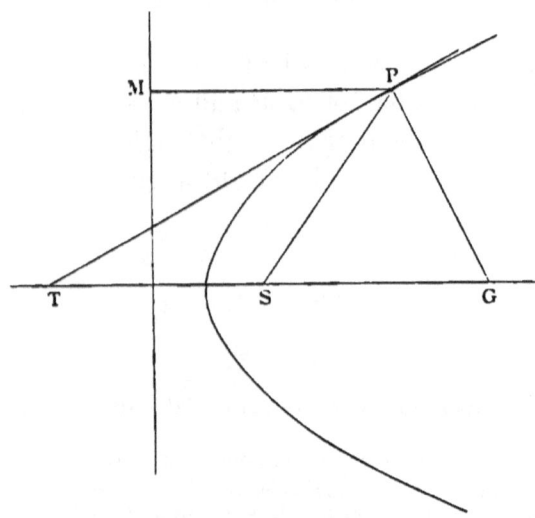

Draw PM perpendicular to the directrix.
Then
$$\angle STP = \angle MPT \quad \text{[Euc. I. 29.}$$
$$= \angle SPT \quad \text{[Prop. 6.}$$
$$\therefore SP = ST.$$

And since TPG is a right angle, a circle centre S and distance SP or ST will pass through G (Euc. III. 31);
$$\therefore SG = SP = ST.$$

1. Prove that SM and PT bisect each other at right angles.
2. If T is the middle point of AX, then N is the middle point of AS.
3. If the triangle SPG is equilateral, the angle TMG is a right angle.
4. A circle can be described round the quadrilateral $SPMZ$, and this circle touches PG at P.
5. If the radius of this circle equal MZ, the triangle SPG is equilateral.
6. The angle between any two tangents to a parabola is half the angle which their chord of contact subtends at the focus.
7. The base AB and the angle C of a triangle ABC are given. Find the locus of the focus of a parabola touching CA, CB in A and B.
8. Two parabolas have the same focus, and their axes in the same straight line, but in opposite directions. Prove that they intersect at right angles.

PARABOLA.

DEF. If the tangent and ordinate at the point P meet the axis in T and N respectively, NT is called the *subtangent* of the point P.

PROPOSITION VIII.

Subtangent $NT = 2AN$.

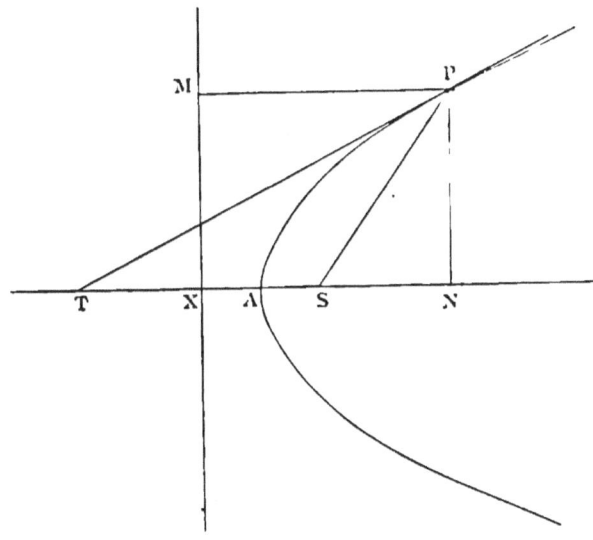

Draw PM perpendicular to the directrix.

Then $\qquad\qquad ST = SP \qquad\qquad$ [Prop. 7.
$\qquad\qquad\qquad = PM$
$\qquad\qquad\qquad = XN;$

and $\qquad\qquad AS = AX,$
$\qquad\qquad \therefore AT = AN;$
$\qquad\qquad \therefore NT = 2AN.$

1. If R be the radius of the circle described round the triangle PNT, prove that $R^2 = SP \cdot AN$.

2. From S a line SQ is drawn parallel to the tangent at P, meeting PE, which is parallel to the axis, in E. Shew that the locus of E is a parabola, vertex S and latus rectum $= \frac{1}{2}$ that of original parabola.

14   PARABOLA.

DEF. If the normal and ordinate at the point P meet the axis in the points G and N respectively, NG is called the *subnormal* of P.

PROPOSITION IX.

Subnormal $NG = 2AS.$

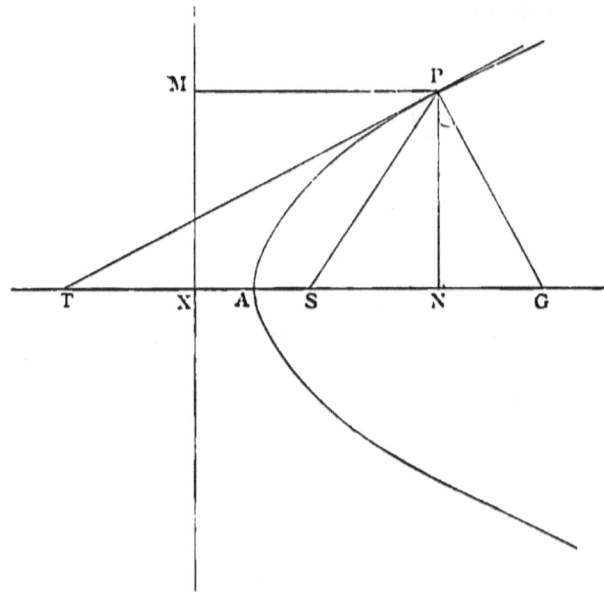

Draw PM perpendicular to the directrix.

Then $SG = SP$ [Prop. 7

$= PM$

$= XN;$

$\therefore NG = SX$

$= 2AS.$

1. If the triangle SPG is equilateral, $SP =$ latus rectum.
2. Deduce Proposition 4 from Propositions 8 and 9.
3. To draw the normal to the curve at any given point.
4. If QM, the ordinate of Q, bisect NG, prove that $QM = PG$.
5. TP, TQ are tangents to a given circle. Construct a parabola which shall touch TP in P and have TQ for axis.

PROPOSITION X.

If the tangent at any point P *intersects the tangent at the vertex in* Y, *then* SY *bisects* PT *at right angles, and is a mean proportional between* SA *and* SP ($SY^2 = AS . SP$).

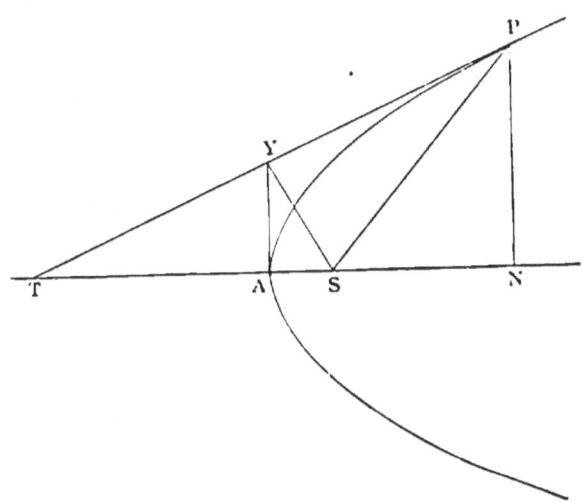

Join SP, and draw PN perpendicular to the axis.

Then, since TN is bisected in A, and AY is parallel to PN,

∴ PT is bisected in Y.

The angles SYT, SYP are equal;   [Euc. I. 8.

∴ SY is at right angles to PT.

Again, because YA is drawn from the right angle perpendicular to the base ST of the triangle SYT,

∴ $SY^2 = SA . ST$ [Euc. VI. 8.

    $= SA . SP$.   [Prop. 7.

1. The circle on SP as diameter touches the tangent at the vertex in Y.
2. Prove $PY . PZ = SP^2$.
3. Prove $PY . YZ = AS . SP$.
4. SY produced meets the directrix in M.
5. If a circle be described on the latus rectum as diameter, and PQ be a common tangent to the parabola and circle, touching them in P and Q respectively, shew that SP, SQ are each inclined to the latus rectum at an angle of $30°$.
6. Given two tangents to a parabola and the focus, shew how to draw the tangent at the vertex, and hence the axis and directrix of the parabola.
7. A long rectangular slip of paper is folded so that one of the corners always lies on the opposite side. Prove that the crease always touches a parabola, of which the opposite side is the directrix.

Proposition XI.

If from any point O *on the tangent at* P, OI *is drawn perpendicular to the directrix, and* OU *perpendicular to* SP, *then* SU = OI. (Adams's property.)

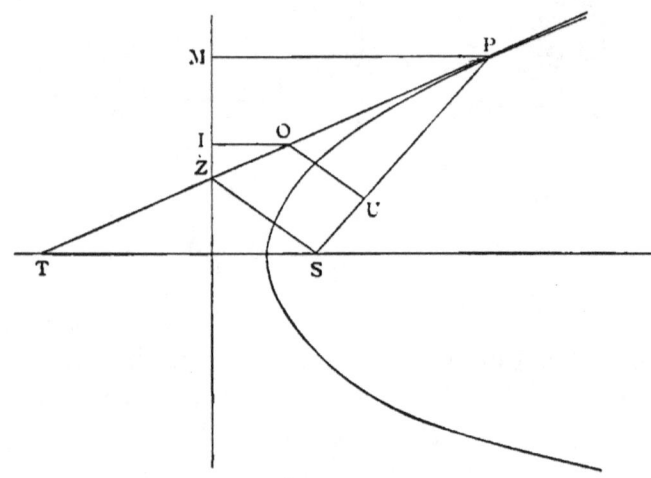

Join SZ, and draw PM perpendicular to the directrix.

Then, since angle ZSP is a right angle,

∴ ZS is parallel to OU.

∴ $SU : SP = ZO : ZP$
$= OI : PM.$

But $SP = PM;$

∴ $SU = OI.$

PARABOLA.

Proposition XII.

To draw two tangents to the parabola from an external point O.

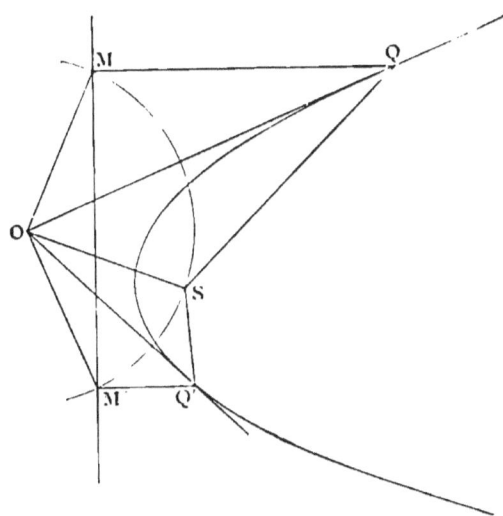

(Analysis.

Let OQ, OQ' be the two tangents. Draw QM, $Q'M'$ perpendiculars on the directrix, and join OS, OM, OM'.

Then, since the angle SQM is bisected by OQ, therefore the triangles SQO, MQO are equal (Euc. i. 4) and $OM = OS$.

So $OM' = OS$. Thus the points M and M' are found, hence construction.)

With centre O at distance OS describe a circle, cutting the directrix in M and M'.

From M and M' draw MQ, $M'Q'$ to the parabola, at right angles to the directrix.

Join OQ, OQ'. OQ, OQ' shall be the tangents required.

Join OS, OM, OM', SQ, SQ'.

Then, in the triangles SQO, MQO,

SQ, $QO = MQ$, QO, and the base OM = base OS;

∴ the angle SQO = angle MQO;

∴ OQ is the tangent at Q.   [Prop. 5.

So OQ' is the tangent at Q'.

Note. The construction may be made on the principles proved in Propositions 10 or 11.

For riders see p. 25.

C. G.

Proposition XIII.

The two tangents OQ, OQ' subtend equal angles at the focus, and the triangles SOQ, SQ'O are similar.

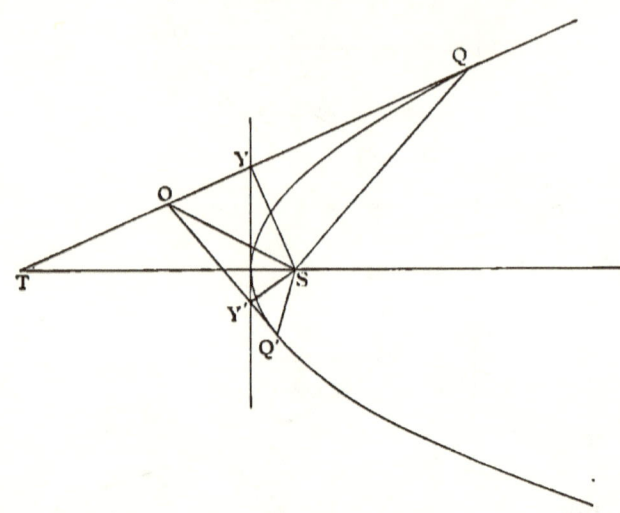

Draw the tangent at the vertex, meeting OQ, OQ' in Y and Y'.

Join SQ, SQ', SY, SY'.

Produce QO to meet the axis in T.

Then, since the angles at Y and Y' are right angles,
[Prop. 10.

the circle on OS as diameter will pass through Y and Y'.

Therefore angle SOQ' = angle SYY' in same segment

$\qquad\qquad\qquad$ = angle STY $\qquad\qquad$ [Euc. vi. 8.

$\qquad\qquad\qquad$ = angle SQO. $\quad$ [Prop. 7 and Euc. i. 5.

Similarly angle $SQ'O$ = angle SOQ;

∴ remaining angles OSQ, OSQ' are equal,

and the triangles SOQ, $SQ'O$ are similar.

OS and a line through O parallel to axis make equal angles with the tangents.

For riders see p. 25.

PARABOLA.

PROPOSITION XIV.

If a pair of tangents OQ, OQ' are drawn to a parabola, and OV is drawn parallel to the axis, meeting QQ' in V, QQ' will be bisected in V.

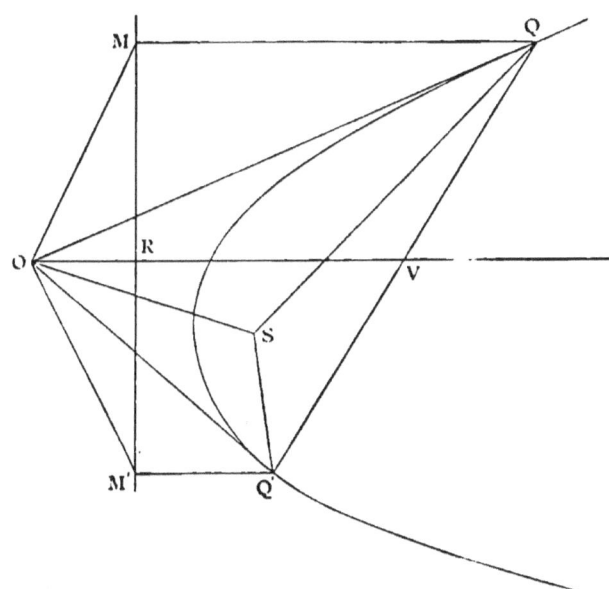

Let OV cut the directrix in R.
Draw QM, $Q'M'$ perpendicular to the directrix.
Join OM, OS, OM', SQ, SQ'.
Then, in the triangles SQO, MQO,
$$SQ, QO = MQ, QO,$$
and angle SQO = angle MQO;   [Prop. 5.
$$\therefore OM = OS.$$
Similarly $\qquad OM' = OS$;
$$\therefore OM = OM',$$
and OR, which is drawn at right angles to the base of the isosceles triangle OMM', bisects it;
$$\therefore MR = M'R.$$
But $\qquad QV : Q'V = MR : M'R$;
$$\therefore QV = Q'V, \text{ or } QQ' \text{ is bisected in } V.$$

For riders see p. 25.

Proposition XV.

The locus of the middle points of any system of parallel chords of a parabola is a straight line parallel to the axis. And the tangent at its point of intersection with the parabola is parallel to the chords.

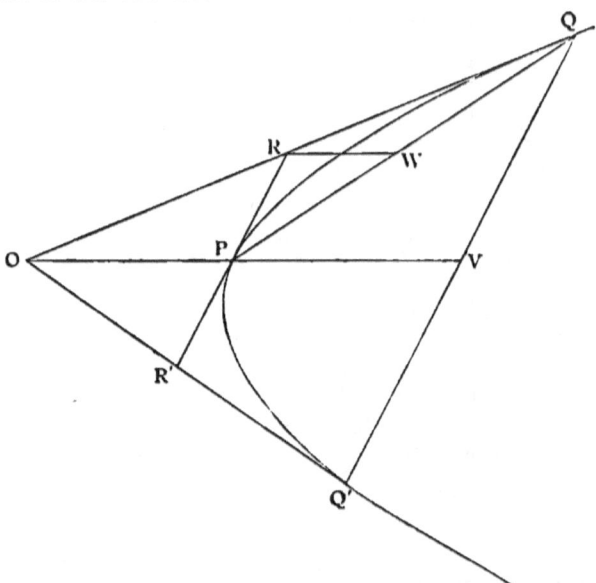

Let QQ' be one of the chords, and RPR' the tangent parallel to them, touching the parabola at a fixed point P.

Through P draw OPV parallel to the axis, meeting QQ' at V and the tangent QRO at O. Join PQ and draw RW parallel to the axis, bisecting PQ at W.   [Prop. 14.

Then $OR = RQ$ because RW is parallel to OP, [Euc. VI. 2. and $OP = PV$ because PR is parallel to QV.

Similarly if we draw a tangent $Q'R'O'$ meeting OPV at O', $O'P = PV$, hence O and O' are coincident.

Since OQ, OQ' are tangents and OV is parallel to axis, QQ' is bisected at V.   [Prop. 14.

Hence the locus of the middle points of all chords parallel to RPR' is a straight line through P parallel to the axis.

DEF. The locus of the middle points of any system of parallel chords drawn in a curve is called a *diameter*.

NOTE. A diameter will not be a straight line for all curves. It has just been proved to be so for a parabola.

For riders see p. 25.

PARABOLA.

Def. The half chords (QV) intercepted between the diameter and the curve are called *ordinates to the diameter*.

Proposition XVI.

If QV is the ordinate of a diameter PV, and the tangent at Q meets VP produced in O, then $OP = PV$.

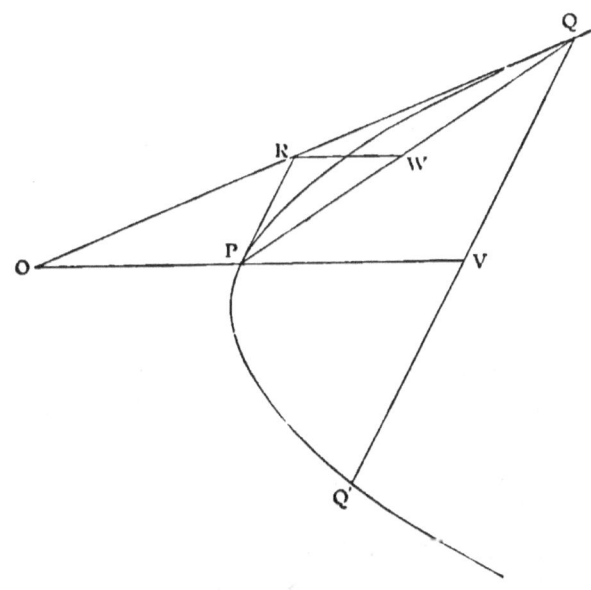

Draw PR touching the parabola at P and meeting OQ at R; through R draw RW parallel to the axis.

Since RP, RQ are a pair of tangents,

PQ is bisected at W,   [Prop. 14.

and PR is parallel to QV;   [Prop. 15.

$$\therefore OP : PV = OR : RQ$$
$$= PW : WQ.$$

But $PW = WQ$, $\therefore OP = PV$.

Proposition XVII.

If QV *is an ordinate to the diameter* PV, *then*
$$QV^2 = 4SP \cdot PV.$$

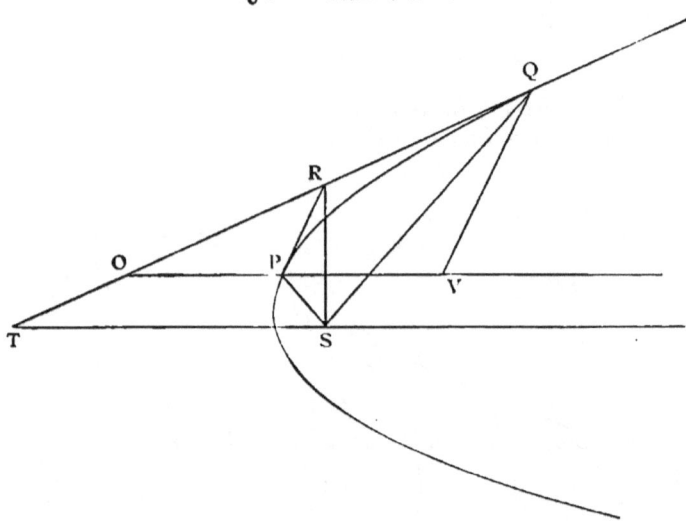

Let the diameter PV meet the parabola in P.

Draw the tangent at Q, meeting the diameter in O and the axis in T.

Draw the tangent at P, meeting OQ in R.

Join SP, SR, SQ.

Then, since RP, RQ are two tangents,

∴ the triangles SRP, SQR are similar ; [Prop. 13.

∴ the angle SRP = angle SQR
   = angle STR [Prop. 7.
   = angle POR, [Euc. I. 29.

and the angle SPR = angle OPR.

∵ the tangent at P bisects the angle SPO, [Prop. 5.

∴ the triangles SRP, POR are similar.

∴ $PR^2 = SP \cdot PO$.

Now OV is bisected in P (Prop. 16), ∴ $QV = 2PR$,

∴ $QV^2 = 4PR^2$
   $= 4SP \cdot PO = 4SP \cdot PV$.

For riders see pp. 25 and 26.

Proposition XVIII

If the focal chord QSQ' is bisected by the diameter PV, which meets the curve in P, QQ' = 4SP.

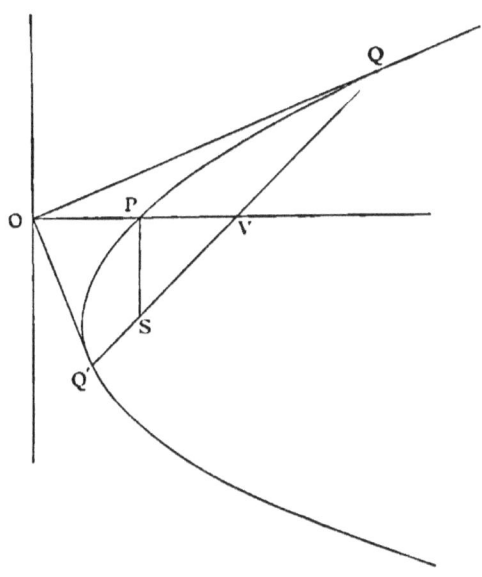

Draw the tangents OQ, OQ' meeting at right angles on the directrix. (Prop. 6.)

Draw the diameter OV. Join SP.

Then, since OV bisects the base of the right-angled triangle QOQ',

$$\therefore QV = OV ; \qquad \text{[Euc. III. 31.}$$
$$\therefore QQ' = 2OV.$$

But $\qquad OP = SP,$ [Def. of parabola.
$$\therefore OV = 2SP ; \qquad \text{[Prop. 16.}$$
$$\therefore QQ' = 4SP.$$

For riders see p. 26.

Proposition XIX.

If two chords, QQ', qq', of a parabola intersect one another, the rectangles contained by their segments are in the ratio of the parallel focal chords; or

$$QO \cdot Q'O : qO \cdot q'O = 4SP : 4Sp.$$

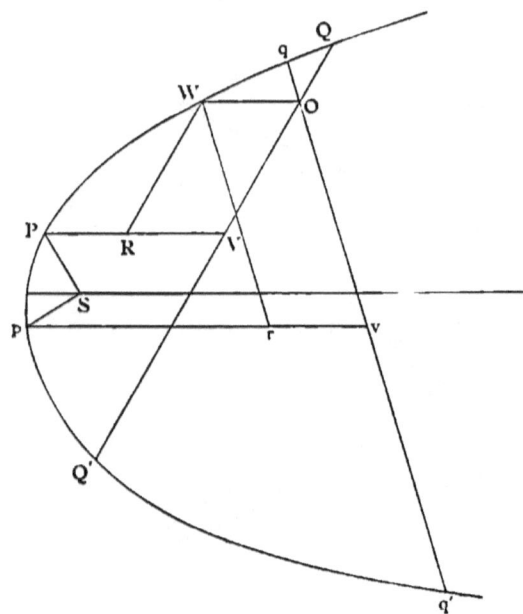

Draw the diameter PV to bisect QQ' in V.

Draw OW parallel to the axis, to meet the parabola in W.

Draw the ordinate WR to the diameter PV. Join SP.

Then
$$QO \cdot Q'O = QV^2 - OV^2 \quad \text{[Euc. II. 5.}$$
$$= QV^2 - WR^2 \quad \text{[Euc. I. 34.}$$
$$= 4SP \cdot PV - 4SP \cdot PR \quad \text{[Prop. 16.}$$
$$= 4SP \cdot RV$$
$$= 4SP \cdot OW.$$

Similarly $\quad qO \cdot q'O = 4Sp \cdot OW;$

$\therefore QO \cdot Q'O : qO \cdot q'O = 4SP : 4Sp.$

For riders see p. 26.

Prop. XII.

1. If the point O be on the directrix, shew from the construction that the tangents intersect at right angles.

2. Find the point O so that the figure $OQSQ'$ may be a parallelogram.

Prop. XIII.

1. If a third tangent be drawn cutting OQ, OQ' in R and T, prove that the circle which circumscribes the triangle ORT will pass through S.

2. What is the locus of the focus of a parabola which touches three given straight lines?

3. A parabola touches each of four straight lines given in position. Give a Geometrical construction for finding its focus.

4. Prove that OS is a mean proportional between OQ and OQ'. What previous proposition is a particular case of this?

5. Two tangents to a parabola and the point of contact of one of them are given. Shew that the locus of the focus is a circle passing through the given point of contact and the intersection of the tangents, and touching one of them.

6. The straight line which bisects the angle QOQ' between the two tangents meets the axis in R. Shew that $SO = SR$.

Prop. XIV.

1. The circle on any focal chord as diameter touches the directrix.

2. The normals at the extremities of a focal chord intersect on the diameter which bisects the chord.

3. Given two tangents and their points of contact, to find the focus and directrix.

Prop. XV.

1. Tangents at the extremities of all parallel chords meet on the same straight line.

2. A parabola being traced on paper, find its axis and directrix.

3. If a chord make an angle of $45°$ with the axis, the line through their middle points passes through an extremity of the latus rectum.

Prop. XVII.

1. If QD be drawn perpendicular to OV, $QD^2 = 4AS \cdot PV$.

2. If TPV is diameter at P, QV an ordinate, and QT tangent at Q, and if $QV = TV$, shew that T is on the directrix.

3. Any chord LVL' is drawn through V, and LM, $L'M'$ are the ordinates of LL' drawn to the diameter PV. Prove that $LM \cdot L'M' = QV^2$.

4. If from the point of contact of a tangent to the parabola a chord be drawn, and another line be drawn parallel to the axis, meeting the tangent, curve, and chord, this line will be divided by them in the same ratio as it divides the chord.

5. Draw a chord of a parabola through a given point, so as to be cut in a given ratio at the point.

Prop. XVIII.

1. To draw a focal chord PSQ such that $SP = 3SQ$.

2. If a diameter meet the directrix in O, OS is perpendicular to the chords bisected by the diameter.

Prop. XIX.

1. The semi latus rectum is a harmonic mean between the segments of any focal chord.

2. If QV be an ordinate to the diameter PV, and pv meeting PQ in v be the diameter conjugate to PQ, then $pv = \frac{1}{4}PV$.

ORTHOGONAL PROJECTIONS.

DEF. I. If from any point a perpendicular be drawn to a fixed plane, the foot of the perpendicular is called the *projection of the point*, and the fixed plane is called the *plane of projection*.

II. *The projection of a line*, straight or curved, is the aggregate of the projections of its points, that is the locus of the feet of perpendiculars, drawn from points on the line, to the plane of projection.

III. *The projection of an area* is the area contained by the projection of the line or lines containing the given area.

IV. The straight line, in which the plane, containing a given curve, intersects the plane of projection, is called the *base line*.

Proposition α.

The projection of a straight line is a straight line.

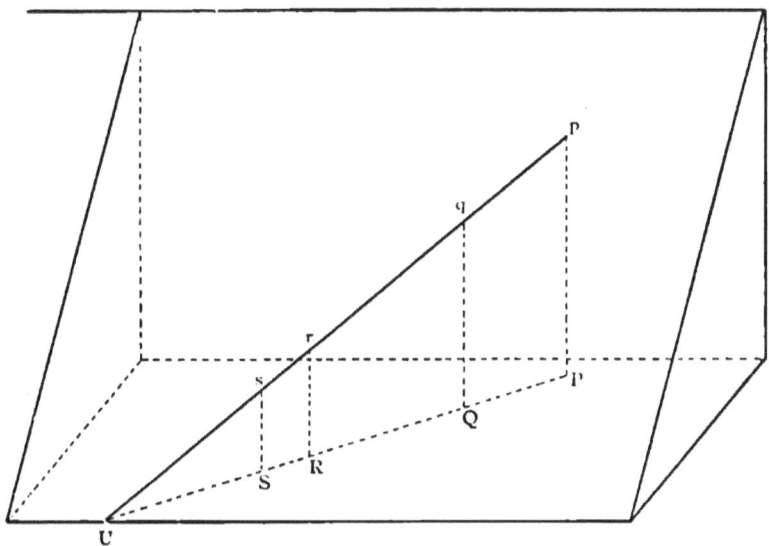

Let *pqrsU* be the given straight line meeting the base line in *U*, and let *P, Q, R, S* be the projections of *p, q, r, s*.

Then the perpendiculars *pP, qQ, rR, sS* will lie in one plane *pPU* (Euc. XI. 6, 7) which intersects the plane of projection in a straight line *UP* (Euc. XI. 3).

Hence the projection of *Up* is the straight line *UP*, and they intersect in a point *U* on the base line.

Proposition β.

The ratio of the segments of a finite straight line is unaltered by projection.

Let *pqrsU* be the given straight line, and *PQRSU* its projection.

Then *pP, qQ, rR, sS* are parallel because they are all perpendicular to the plane of projection, and they are all in the same plane *PUp*; hence the segments *PQ, QR, RS* are in the same ratio as *pq, qr, rs* (Euc. VI. 2).

Proposition γ.

Parallel straight lines project into parallel straight lines of proportional length.

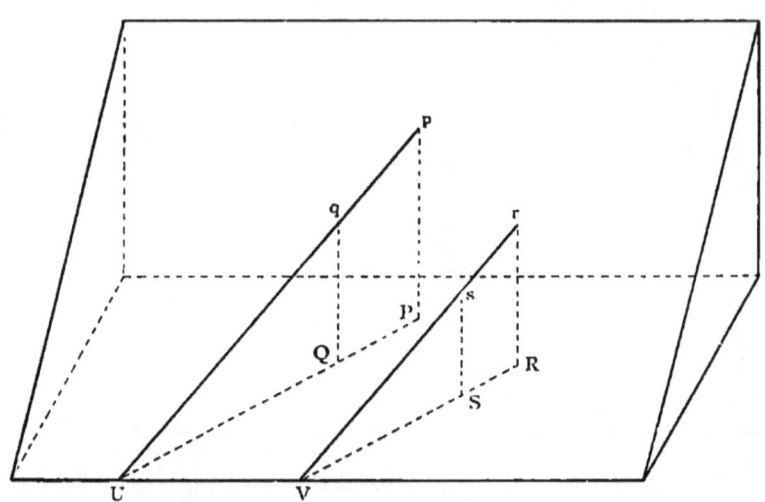

Let pqU, rsV be two parallel straight lines, meeting the base line in U and V, and let PQU, RSV be their projections.

pP and rR are parallel,   [Euc. XI. 6.

pq and rs are parallel;   [hyp.

∴ the plane UpP is parallel to plane VrR. [Euc. XI. 15.

Hence PQU is parallel to RSV.   [Euc. XI. 16.

Again, triangles pUP, rVR are equiangular, [Euc. XI. 10.

$$\therefore PQ : pq = PU : pU,$$
$$= RV : rV,$$
$$= RS : rs.$$

Obs.—This ratio $PU : pU = \cos pUP$.

Proposition δ.

A tangent projects into a tangent, cutting the base line in the same point.

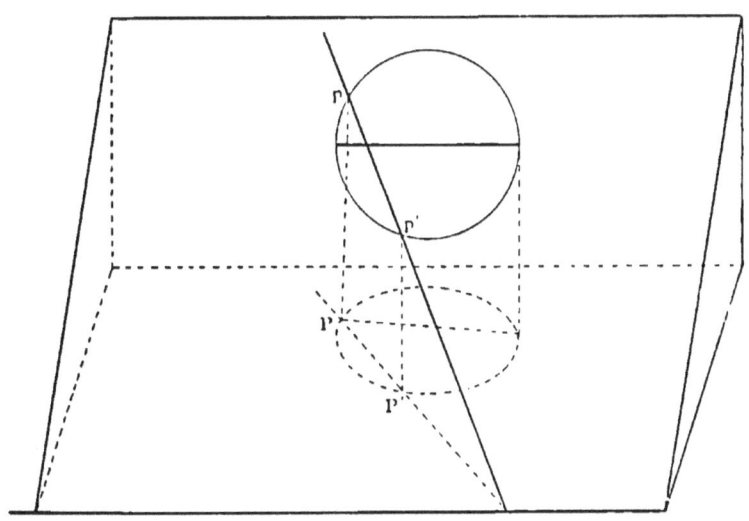

Let pp' be two points on a curve near to one another, then their projections PP' lie on the projection of the given curve.

Let p' move up to and coincide with p, so that pp' becomes a tangent to the given curve.

Then P' moves up to and coincides with P, and PP' becomes a tangent to the projection of the given curve.

Also these straight lines meet the base line in the same point. (Prop. α)

Proposition ε.

The ratio of areas is unaltered by projection.

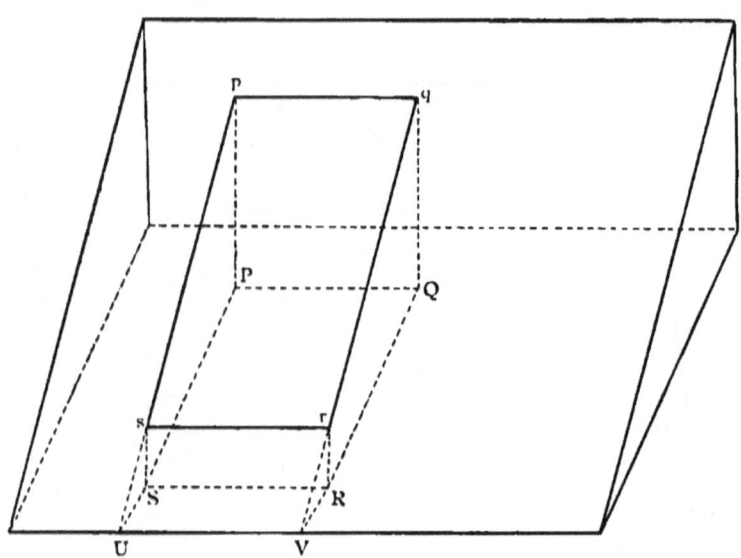

Case 1. Let $pqrs$ be a rectangle, having two sides pq, rs parallel to the base line, and let $PQRS$ be its projection; produce ps, qr to meet the base line in U, V.

Area $PQRS$: area $pqrs = PQ \times PS : pq \times ps$,

$$= PS : ps,$$
$$= PU : pU.$$

Now this ratio (which is equal to $\cos \alpha$, if α be the angle between the original plane and the plane of projection) is independent of the length and breadth of the rectangle; therefore all such rectangles are diminished by projection in the same proportion, and all such rectangles drawn in the original plane bear the same ratio to one another as their projections do.

ORTHOGONAL PROJECTIONS. 31

Case 2. But a figure of any shape may be divided into a large number of narrow strips by lines perpendicular to

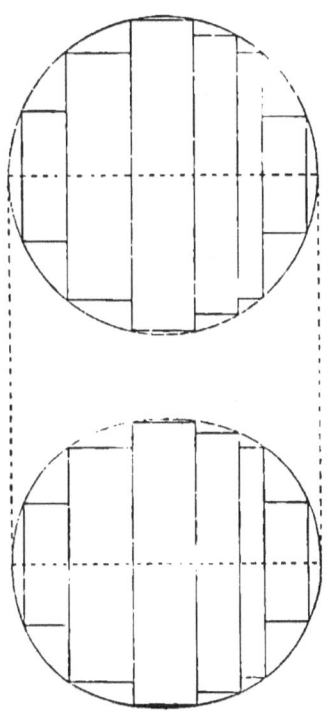

the base line, and each of these strips will form one of these rectangles, with two small areas at each end; now the sum of these rectangles bears to the sum of their projections a constant ratio, also by increasing the number of rectangles and decreasing their width the difference between them and the given area may be indefinitely diminished, hence an area of any shape is diminished by projection in the same ratio (1 : cos α) and all areas in the original plane bear the same ratio to one another as their projections do.

Proposition ζ.

The projections of two straight lines at right angles to one another are lines at right angles to one another, if one of the original lines is parallel to the base line.

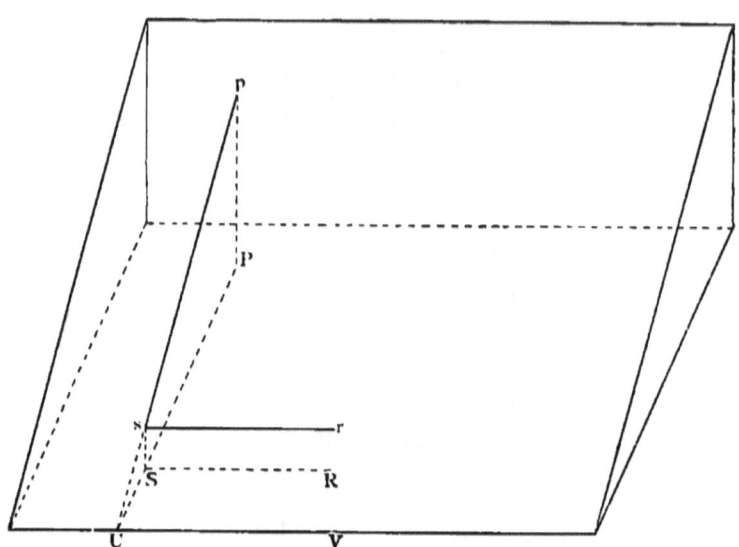

Let ps, sr be two straight lines at right angles to one another, of which sr is parallel to the base line UV. Let PS, SR be their projections. Since sr is parallel to UV, it does not meet the plane of projection $PSUV$, hence sr does not meet SR; also sr, SR are in the same plane, therefore they are parallel to one another.

But SR is at right angles to Ss,

therefore sr is at right angles to Ss; [Euc. I. 29.

also sr is at right angles to ps, [hyp.

∴ sr is at right angles to the plane $psUSP$; [Euc. XI. 4.

∴ SR is at right angles to the plane $psUSP$, [Euc. XI. 8.

and PSR is a right angle.

NOTE. The projection of a right angle is *not* a right angle, unless one of the arms of the original angle is parallel to the base line.

ELLIPSE.

DEF. I. An *ellipse* is the locus of a point (P) whose distance from a fixed point (S) bears a constant ratio (e), less than unity, to its distance (PM) from a fixed straight line (XM),

$$(SP = e \cdot PM).$$

II. The fixed point (S) is called the *focus*.

III. The fixed straight line (XM) is called the *directrix*.

IV. The constant ratio (e) is called the *eccentricity*.

Proposition I.

Construction for points on the ellipse.

The perpendicular on the directrix through the focus is an axis of symmetry.

To find the vertices A *and* A'.

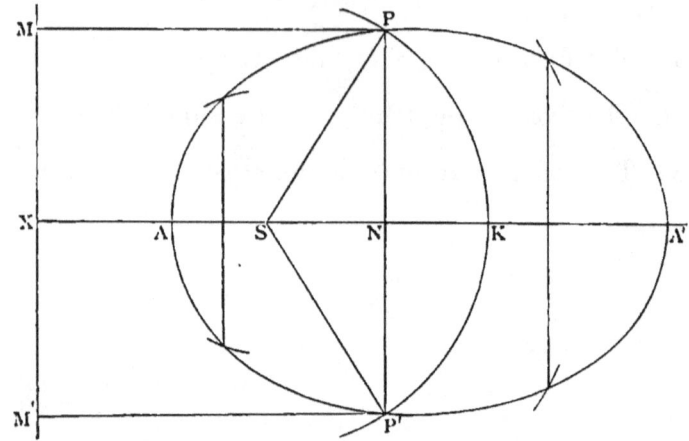

From the focus S draw SX perpendicular to the directrix. Divide XS in A, so that
$$SA = e \cdot AX;$$
also in XS produced take A' so that
$$SA' = e \cdot A'X.$$
Then A and A' are points on the curve.

Take any point N on the straight line AA', with centre S and radius $e \cdot XN$ describe a circle; through N draw PNP' perpendicular to AA' and cutting the circle in P and P', then P and P' are points on the ellipse. Draw PM, $P'M'$ perpendicular to the directrix,
$$SP = e \cdot XN = e \cdot PM,$$
$$SP' = e \cdot XN = e \cdot P'M'.$$

Corresponding to any point N on the line AA', we thus get two points P and P' at equal distances on opposite sides of AA'; hence the ellipse is symmetrical with respect to AA', or AA' is an axis, and the points A and A' are vertices.

NOTE. It may be proved that the circle intersects the perpendicular NP, when N is any part of the axis AA' between A and A', but not when N lies outside the part AA', hence the ellipse lies entirely between lines drawn through A and A' at right angles to the axis. See Appendix.

For riders see p. 37.

Proposition II.

If the chord PP′ *intersects the directrix in* K, SK *bisects the exterior angle between* SP *and* SP′.

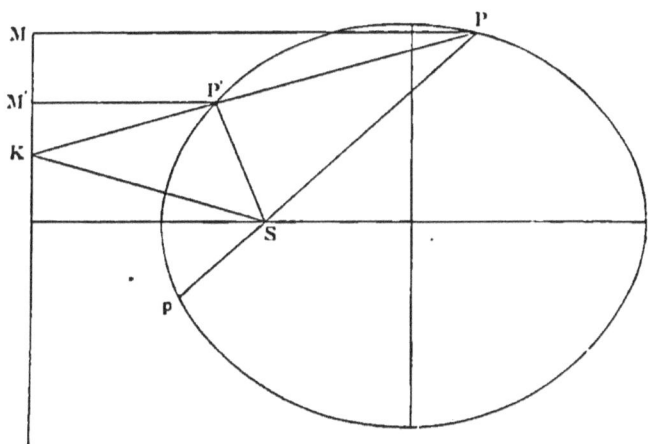

Join SP, SP', SK; produce PS to p, and draw PM, $P'M'$ perpendicular to the directrix.

Then $\qquad SP = e \cdot PM,$
and $\qquad SP' = e \cdot P'M'$;
$$\therefore SP : SP' = PM : P'M'$$
$$= PK : P'K,$$
by similar triangles PKM, $P'KM'$.

Therefore SK bisects $P'Sp$ (Euc. vi. A.).

Prop. II.

1. PSP_1 is a focal chord. Prove that XP and XP_1 are equally inclined to the axis.

2. PSP_1 is a focal chord. PA, P_1A are produced to meet the directrix in K and K_1 respectively. Prove that KSK_1 is a right angle.

3. Two chords PQ, $P'Q$ meet the directrix in p, p' respectively. Prove that the angle pSp' is half the angle PSP'.

4. If the focus of an ellipse and two points on the curve be given, the directrix will pass through a fixed point.

ELLIPSE.

DEF. If the axis through the focus (S) meets the ellipse at A and A', AA' is called the *major axis*.

DEF. Bisect AA' in C, then C is called the *centre of the ellipse*.

DEF. The double ordinate BCB', drawn through C, is called the *minor axis*.

PROPOSITION III.

If PN *is the ordinate of a point* P *on the ellipse,*
$$PN^2 : AN \cdot A'N = CB^2 : CA^2,$$
and CB *is less than* CA.

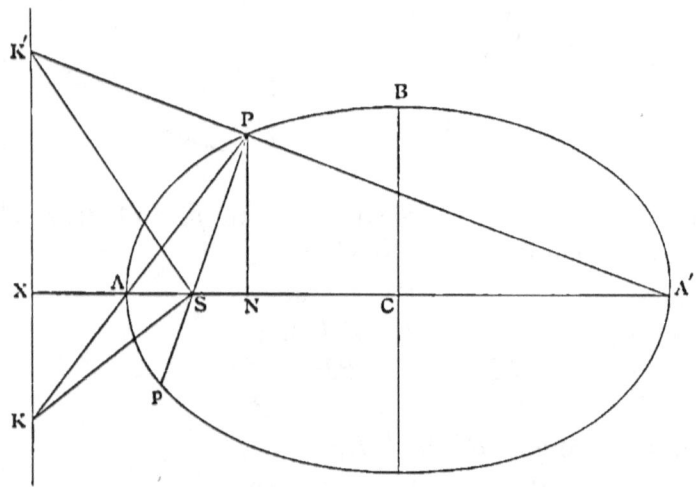

Join PA, $A'P$, and produce them to meet the directrix at K and K'.

Join SP, SK, SK', and produce PS to p.

By similar triangles PAN, KAX,
$$PN : AN = KX : AX.$$
By similar triangles $PA'N$, $K'A'X$,
$$PN : A'N = K'X : A'X;$$
$$\therefore PN^2 : AN \cdot A'N = KX \cdot K'X : AX \cdot A'X.$$

ELLIPSE. 37

But SK bisects the angle ASp,   [Prop. 2.
and SK' bisects the angle ASP,   [Prop. 2.

$\therefore KSK'$ is a right angle;

$\therefore KX . K'X = SX^2$;   [Euc. VI. 8.

$\therefore PN^2 : AN . A'N = SX^2 : AX . A'X$.

Similarly, since P may coincide with B,

$$BC^2 : AC^2 = SX^2 : AX . A'X,$$
$$\therefore PN^2 : AN . A'N = BC^2 : AC^2.$$

Again, $BC^2 : AC^2 = SX^2 : AX . A'X$.

Now $SX = AX + SA = AX(1+e)$,

also $SX = A'X - SA' = A'X(1-e)$,

$\therefore SX^2 = (1-e^2) AX . A'X < AX . A'X$;

$\therefore BC < AC$.

Prop. I.

1. If a parabola and an ellipse have the same focus and directrix, the parabola lies entirely outside the ellipse.

2. A point P lies within, on, or without the ellipse, according as the ratio $SP : PM$ is less than, equal to, or greater than the excentricity, PM being the perpendicular on the directrix.

3. Any chord PQ of an ellipse meets the directrix in R. Prove that
$$SP : PR = SQ : QR.$$

4. A straight line meets the ellipse in P, and the directrix in R. From K, any point in PR, KU is drawn parallel to SR, to meet SP in U, and KI perpendicular to the directrix. Prove that $SU = e . KI$.

Prop. III.

1. If PM be drawn perpendicular to BCB', prove that
$$PM^2 : BM . B'M = CA^2 : CB^2.$$

2. P, Q are two points on an ellipse. $AQ, A'Q$ cut PN or PN produced in L and M. Prove that $PN^2 = LN . MN$.

Proposition IV.

If the ordinates of the circle described on AA′ *as diameter be reduced in the ratio of* CA : CB, *the locus of their extremities is the ellipse.*

$$(PN : pN = CB : CA).$$

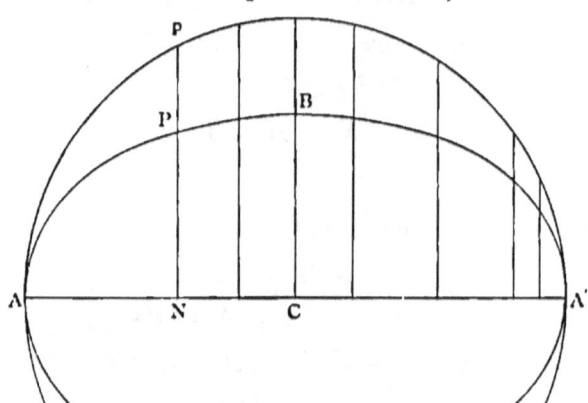

Let ApA' be the circle described on AA' as diameter, and NPp the ordinate of p, meeting the ellipse at P.

$$PN^2 : AN . A'N = CB^2 : CA^2. \quad \text{[Prop. 3.}$$

But $\quad pN^2 = AN . A'N;\quad$ [Euc. III. 3 and 35.

$$\therefore PN^2 : pN^2 = CB^2 : CA^2,$$
$$PN : pN = CB : CA. \qquad \text{q.e.d.}$$

Def. I. The circle described on AA' as diameter is called the *auxiliary circle*.

II. The points p and P lying on a common ordinate of the ellipse and auxiliary circle are called *corresponding points*.

III. A chord of the ellipse and a chord of the auxiliary circle are called *corresponding chords*, if their extremities are corresponding points.

Proposition V.

The projection of a circle is an ellipse.

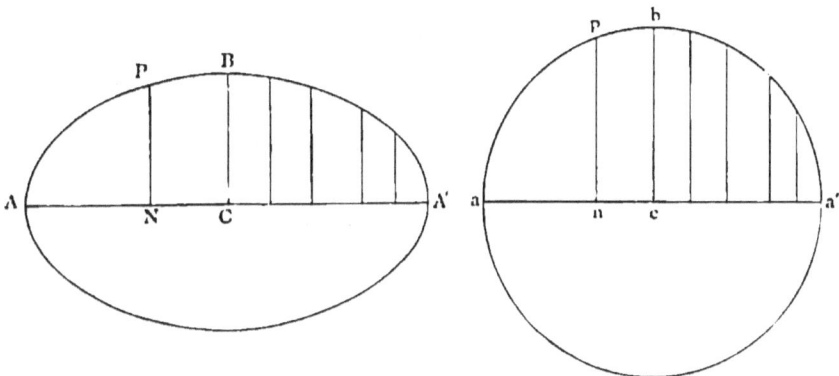

Let apa' be a circle, having its diameter aa' parallel to the base line, cb the radius perpendicular to aa', pn a perpendicular from any point p to aa'.

Let $APBA'$ be the projection of the circle $apba'$, and let the points A, A', B, C, P, N be the projections of the points a, a', b, c, p, n.

Then $pn^2 = an \cdot na'$; [Euc. III. 3 and 35.

$\therefore pn^2 : cb^2 = an \cdot na' : ca^2$.

But $pn^2 : cb^2 = PN^2 : CB^2$, [Prop. γ.

and $an \cdot na' : ca^2 = AN \cdot NA' : CA^2$;

$\therefore PN^2 : CB^2 = AN \cdot NA' : CA^2$.

Also PN and CB are perpendicular to AA'; [Prop. ζ. therefore the locus of P is an ellipse whose axes are CA, CB.
[Prop. 3.

NOTE. The circle aba' is equal to the auxiliary circle. The ratio $CB : CA = \cos a$, where a is the angle of projection.

The area of the ellipse $= \pi AC \cdot BC$.

Proposition VI.

The ellipse is symmetrical with respect to the minor axis, and has a second focus (S') and directrix.

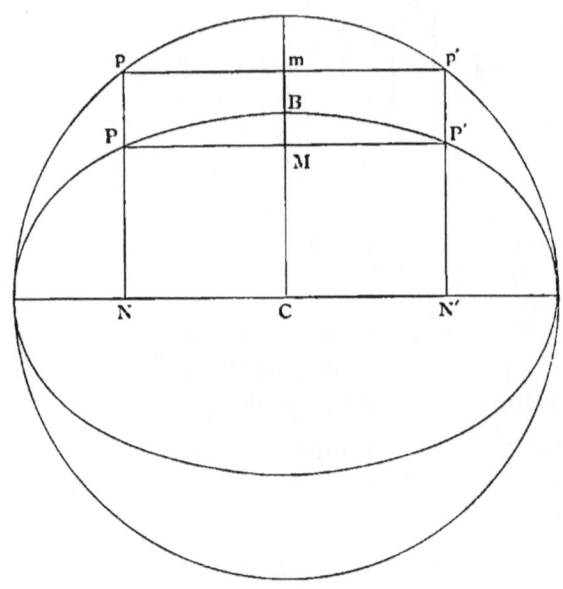

Let pmp' be a chord of the auxiliary circle, cutting the minor axis at right angles in m. Take P and P' points on the ellipse corresponding to p and p', and draw the common ordinates pPN, $p'P'N'$, and join PP', cutting the minor axis in M.

Then $\qquad pN = p'N';$ \qquad [Euc. I. 34.

$\therefore PN = P'N';$ \qquad [Prop. 4.

therefore PP' is parallel to NN' and perpendicular to CB.

Also, $\qquad pm = p'm;$ \qquad [Euc. III. 3.

$\therefore PM = P'M.$ \qquad [Euc. I. 34.

ELLIPSE. 41

Hence, corresponding to any point P on the ellipse, there is another point P' on the ellipse such that the chord PP' is bisected at right angles by the minor axis, or the ellipse is symmetrical with respect to the minor axis.

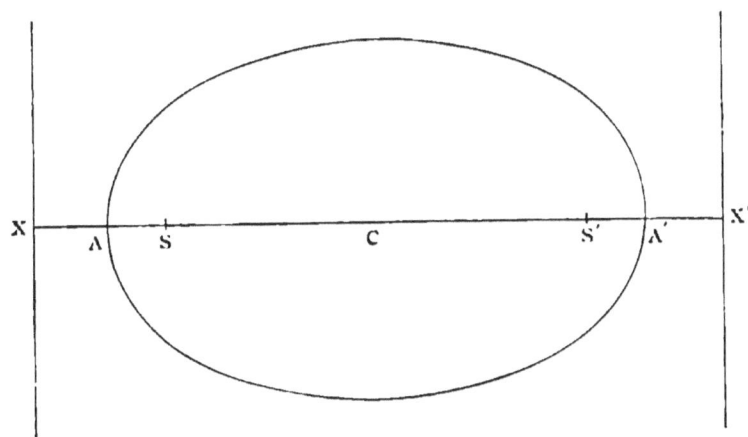

If we take CS' equal to CS, and CX' equal to CX, and through X' draw a line perpendicular to AA', the ellipse can be described with this line as directrix, S' as focus, and eccentricity the same as before.

Prop. IV.

1. A straight line cannot meet the ellipse in more than two points.
2. Of all lines drawn from the centre to the curve CA is the greatest and CB the least.
3. P and Q are corresponding points on the ellipse and the auxiliary circle; through P KPL is drawn making the same angle with the axes which CQ does, and cutting them in K and L. Shew that KL is a constant length.
4. PM drawn perpendicular to BB' meets the circle on the minor axis as diameter in p'. Prove
$$PM : p'M = CA : CB.$$
5. If the two extremities of a rod slide along two fixed straight lines at right angles to one another, any fixed point in the rod will describe an ellipse.

Prop. V.

An ellipse may also be itself projected into a circle.

Proposition VI. (*Aliter.*)

Let aba' be a circle, and ABA' its projection.

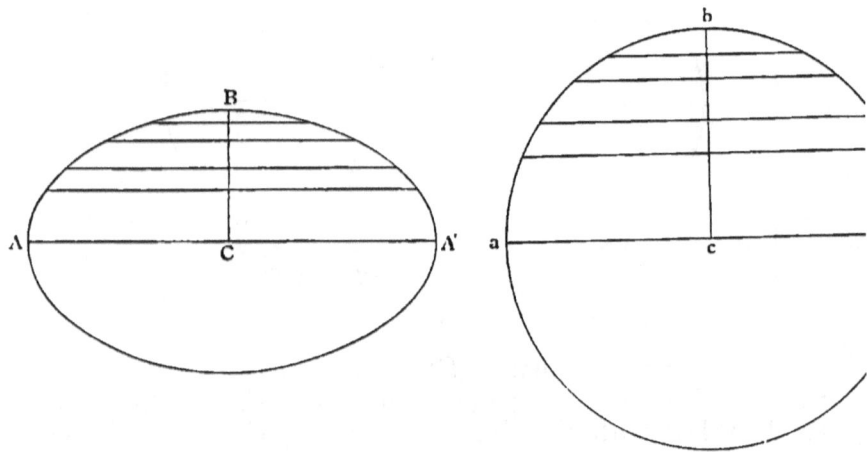

All chords of the circle parallel to aa' are bisected by cb. [Euc. III. 3.

Therefore all chords of the ellipse parallel to AA' are bisected by CB. [Prop. γ.

And CB is perpendicular to chords it bisects. [Prop. ζ.

Hence the ellipse is symmetrical with respect to the minor axis.

And it may be described with reference to a second focus and directrix on the opposite side of the centre.

ELLIPSE.

Proposition VII.

$$CA = e \cdot CX; \quad CS = e \cdot CA; \quad CS \cdot CX = CA^2.$$

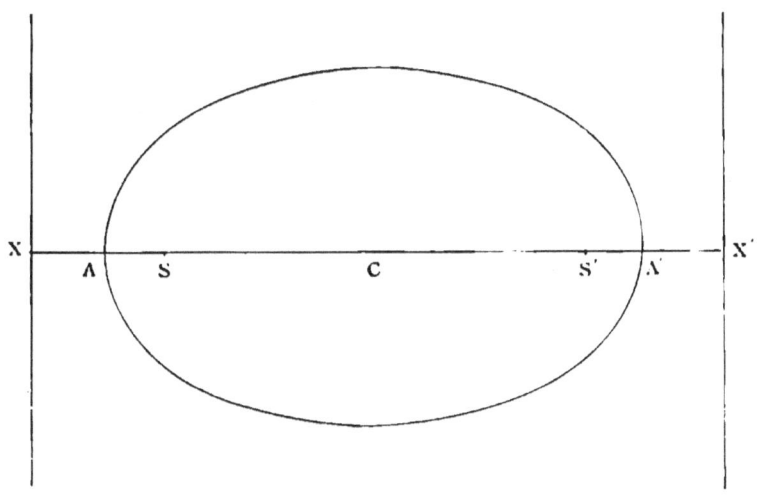

$$SA = e \cdot AX, \qquad \text{[Def.}$$
$$SA' = e \cdot A'X. \qquad \text{[Def.}$$

By addition
$$AA' = e(AX + A'X) = e(AX + AX') = eXX';$$
$$\therefore CA = e \cdot CX \quad \ldots\ldots\ldots\ldots(\alpha).$$

By subtraction
$$SS' = e \cdot AA';$$
$$\therefore CS = e \cdot CA \quad \ldots\ldots\ldots\ldots(\beta);$$
$$\therefore CS \cdot CX = CA^2 \quad \ldots\ldots\ldots\ldots(\gamma).$$

Prop. VII.

Given an ellipse and one focus, find the centre and the eccentricity.

Proposition VIII.

$$SP + S'P = AA'.$$

Mechanical construction for the ellipse.

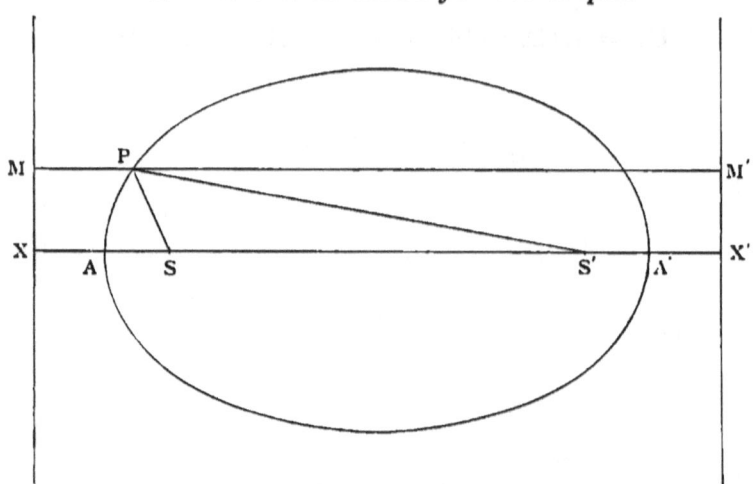

Draw MPM' perpendicular to the directrices.

Then $SP = e \cdot PM$,

and $S'P = e \cdot PM'$;

$\therefore SP + S'P = e \cdot MM'$
$= e \cdot XX'$
$= AA'.$

If an endless string be placed round two drawing-pins at S and S', and kept tight by a pencil point at P, the pencil can be made to trace out an ellipse of which S, S' are the foci.

Prop. VIII.

1. If P be any point, $SP + S'P$ is greater than, equal to, or less than AA', according as P is without, upon, or within the ellipse.

2. A circle is drawn entirely within another circle. Prove that the locus of a point equidistant from the circumferences of these two circles is an ellipse.

3. Two ellipses have a common focus, and their major axes equal. Prove that they cannot intersect in more than two points.

4. Prove that the straight line, which bisects the exterior angle between PS and PS', cannot meet the ellipse again.

Proposition IX.

$$CB^2 = CA^2 - CS^2 = SA \cdot SA'.$$

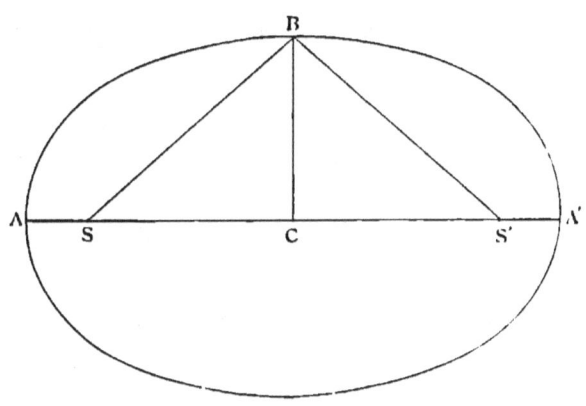

	$SB + S'B = AA'.$	[Prop. 8.
But	$SB = S'B;$	[Euc. I. 4.
	$\therefore SB = CA,$	
	$CB^2 = SB^2 - CS^2$	[Euc. I. 47.
	$= CA^2 - CS^2$	
	$= SA \cdot SA'.$	[Euc. II. 5.

46   ELLIPSE.

DEF. The double ordinate through the focus is called the *latus rectum* (LL').

PROPOSITION X.

The semi latus rectum SL *is a third proportional to* CA *and* CB.

$$SL \cdot CA = CB^2,$$

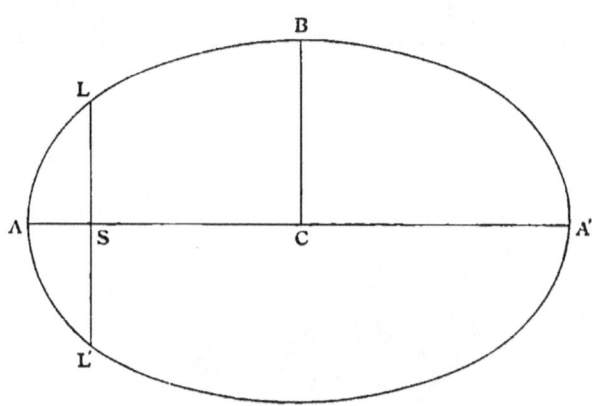

$$SL^2 : AS \cdot A'S = CB^2 : CA^2. \qquad \text{[Prop. 3.}$$
But $$AS \cdot A'S = CB^2; \qquad \text{[Prop. 9.}$$
$$\therefore SL^2 : CB^2 = CB^2 : CA^2;$$
$$\therefore SL : CB = CB : CA;$$
$$\therefore SL \cdot CA = CB^2.$$

Proposition XI.

If the tangent at P meets the directrix in Z, PSZ is a right angle.

Also tangents at the ends of a focal chord intersect on the directrix.

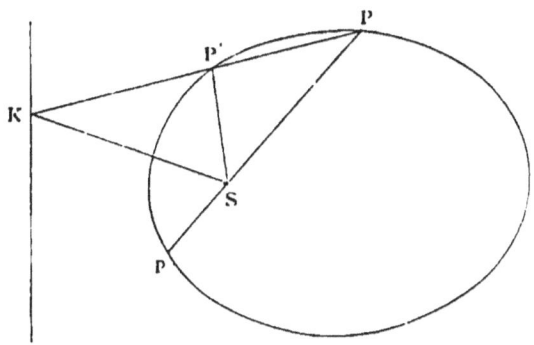

Take a point P' on the ellipse near to P, and let the chord PP' meet the directrix in K, and produce PS to p. Then KS bisects the angle $P'Sp$.   [Prop. 2.

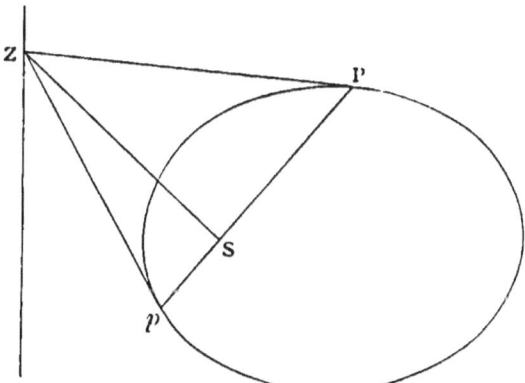

When P' coincides with P, so that $PP'K$ becomes the tangent PZ, $P'Sp$ becomes two right angles; therefore PSZ is a right angle.

Hence ZSp is a right angle, and Zp is the tangent at p, or the tangents at P and p intersect on the directrix.

1. Tangents at the extremities of the latus rectum intersect in X.

2. If through any point P of an ellipse QPN be drawn perpendicular to the axis, meeting the tangent at L in Q and axis in N, $QN = SP$.

3. To draw the tangent at a given point P of the ellipse.

4. By drawing the tangent at B, prove $CS . CX = CA^2$.

Proposition XII.

If the normal at P intersects the major axis in G,
$$SG = e \cdot SP.$$

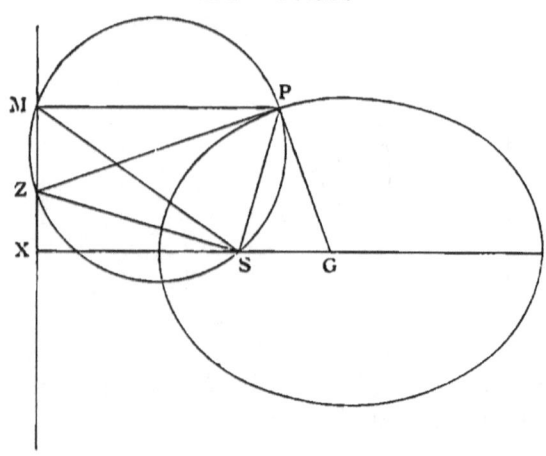

Draw the tangent PZ, join SZ, draw PM perpendicular to the directrix, and join SM.

ZMP and ZSP are right angles; [Prop. 11.
therefore the circle, on ZP as diameter, passes through M and S. [Euc. III. 31.

Since ZPG is a right angle, PG touches the circle.
[Euc. III. 16.

Therefore the angle SPG = angle SMP in the alternate segment. [Euc. III. 32.

Also angle PSG = angle SPM. [Euc. I. 29.

Therefore the triangles SPG, SMP are similar;
$$\therefore SG : SP = SP : PM;$$
$$\therefore SG = e \cdot SP.$$

Prop. XII.

1. P is any point on the ellipse, M a fixed point on the major axis. A perpendicular is drawn from M on the tangent at P. Find the locus of the intersection of this perpendicular with the radius vector SP.

2. If GL be drawn perpendicular to SP, the ratio $PN : GL$ is constant, and PL = semi latus rectum.

3. If PG be produced to meet the minor axis in g, gS produced meets the directrix in M, the foot of the perpendicular from P.

ELLIPSE.

PROPOSITION XIII.

The tangent and normal to an ellipse at any point P are respectively the external and internal bisectors of the angle between the focal distances.

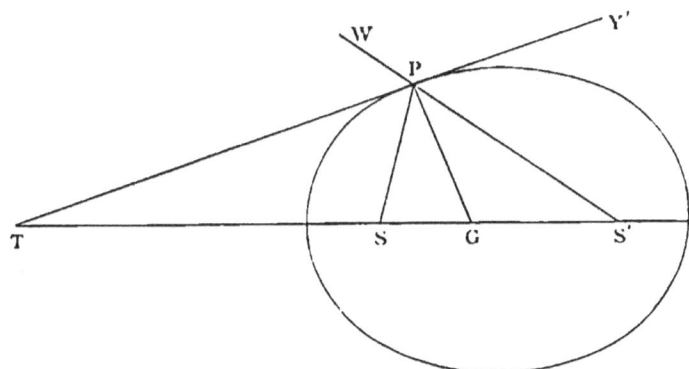

Let TPY' be the tangent and PG the normal,

$$SG = e \cdot SP, \qquad \text{[Prop. 12.}$$

and $$S'G = e \cdot S'P;$$

$$\therefore SG : S'G = SP : S'P;$$

therefore PG bisects the angle SPS'. [Euc. VI. 3.

Therefore the complements SPT, $S'PY'$ are equal, but

$$S'PY' = WPT; \qquad \text{[Euc. I. 15.}$$

therefore PT bisects the exterior angle SPW.

PROP. XIII.

1. If SY, the perpendicular on the tangent at P, meet $S'P$ produced in s, prove (1) $sY = SY$, (2) $SP = Ps$, (3) $S's = AA'$.
If P move round the ellipse what is the locus of s?
[NOTE. On account of (1) s is called the image of the focus in the tangent.]

2. If the tangent and normal meet the minor axis in t and g respectively, the circle on gt as diameter passes through P and the two foci.

3. If the normal at P meet the major and minor axes in G and g, prove that the triangles SPG, gPS' are similar.

4. $\qquad\qquad SP \cdot S'P = PG \cdot Pg.$

5. No normal can pass through the centre, except the normals at the ends of the axes.

6. If a circle be described through the foci of an ellipse, a straight line drawn from its intersection with the minor axis to its intersection with the ellipse will touch the ellipse.

ELLIPSE.

PROPOSITION XIV.

The feet of the perpendiculars (SY, S'Y') *from the foci on the tangent at* P *are on the auxiliary circle.*

Also if CE, *parallel to the tangent at* P, *intersects* S'P *in* E, PE = CA.

Also $\qquad$ SY . S'Y' = CB².

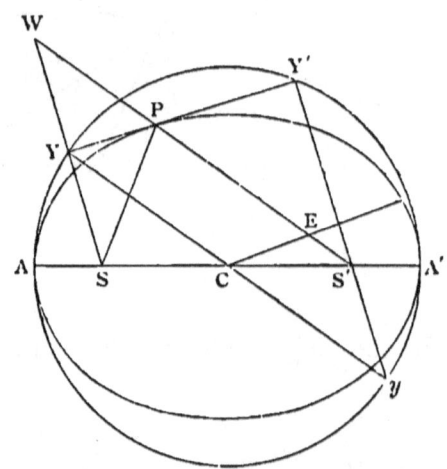

Produce S'P, SY to meet in W. Join CY.

In the triangles YPS, YPW, YP is common, right angles PYS, PYW are equal, angle YPS = angle YPW; [Prop. 13.
∴ SP = PW, SY = YW; [Euc. I. 26.
and SC = CS', ∴ S'W is parallel to CY; [Euc. VI. 2.
∴ CY = ½ S'W [Euc. VI. 4.
  = ½ (S'P + PS) = ½ AA' [Prop. 8.
  = CA ;

therefore Y is on the auxiliary circle.

Similarly, Y' is on the auxiliary circle.

Also YCEP is a parallelogram; therefore
$$PE = CY = CA.$$

Produce Y'S' to meet the circle in y and join Yy.

Then, YY'y being a right angle, Yy passes through the centre C, [Euc. III. 31.
$\qquad$ SY = S'y, [Euc. I. 4.
SY . S'Y' = S'y . S'Y' = AS' . S'A' [Euc. III. 35.
  = CB². [Prop. 9.

For riders see page 52.

Proposition XV.

Corresponding chords of the ellipse and auxiliary circle intersect on the major axis.

Also tangents at corresponding points intersect on the major axis.

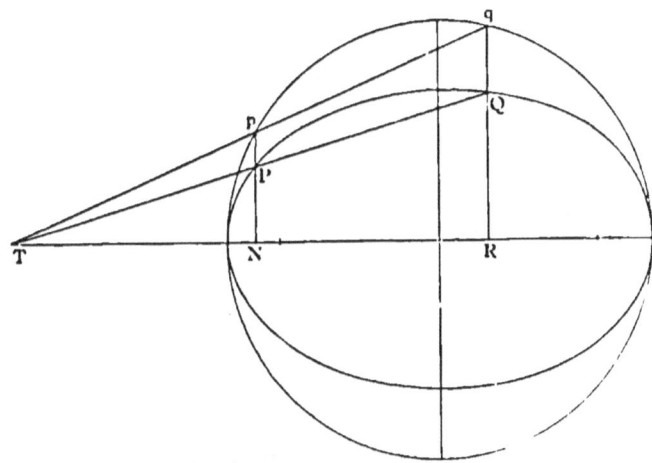

Let PQ be a chord of an ellipse, meeting the major axis in T.

Let p be the point of the auxiliary circle corresponding to P. Join Tp, and produce it to meet the ordinate RQ produced in q.

Then
$$qR : pN = RT : NT \quad \text{[Euc. vi. 4.}$$
$$= QR : PN \quad \text{[Euc. vi. 4.}$$
$$\therefore qR : QR = pN : PN$$
$$= AC : BC; \quad \text{[Prop. 4.}$$

$\therefore q$ is the corresponding point to Q, and the corresponding chords PQ, pq meet the axis in the same point T.

If Q moves up to and coincides with P, then q moves up to and coincides with p, and PT, pT become tangents to the ellipse and circle, or the tangents at corresponding points intersect on the major axis.

Prop. XV.

1. Pp are corresponding points. The tangent at p meets CB produced in K. Prove $CK \cdot PN = AC \cdot BC$.

2. OQ, OQ' are tangents to an ellipse. ON is drawn perpendicular to the axis. Prove that the tangents to the auxiliary circle at the corresponding points q and q' meet in ON.

Prove also that if QQ' produced meet the major axis in T, $CN \cdot CT = CA^2$.

4—2

Proposition XVI.

If the tangent at P *meets the major axis produced at* T,
$$CN \cdot CT = CA^2.$$

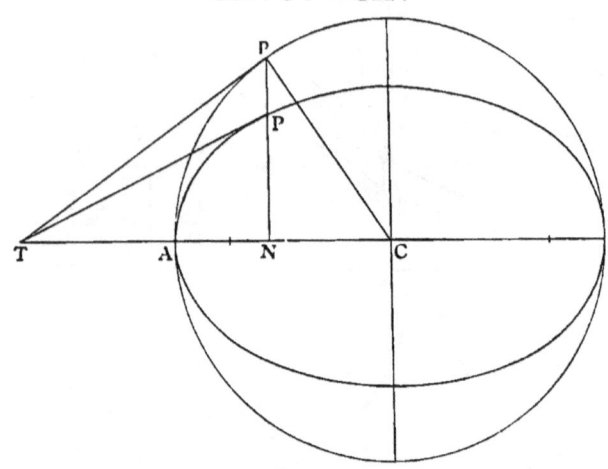

Produce NP to meet the auxiliary circle in p, and join pT, pC.

pT touches the circle;   [Prop. XV.
therefore CpT is a right angle;   [Euc. III. 18.
$\therefore CN \cdot CT = Cp^2$ [Euc. VI. 8.
$= CA^2.$

Prop. XIV.

1. To draw a tangent to the ellipse parallel to a given straight line.
2. If a straight line through C parallel to the tangent intersect the SP, $S'P$ distances in E, E', prove $CE = CE'$.
3. Prove also $S'E = SE'$.
4. The circle described on SP as diameter touches the auxiliary circle.
5. SK is parallel to $S'P$, and YK perpendicular to SK. Shew that the parabola having S for focus and K for vertex touches the ellipse.
6. Given in position a focus and tangent, and in magnitude the minor axis, find the locus of the other focus.
7. A chord of a circle which subtends a right angle at a fixed point envelopes a conic whose foci are the fixed point and the centre of the circle.
8. If a second tangent intersect YPY'' at right angles in O, prove that
$$OY \cdot OY'' = BC^2.$$
Hence prove $CO^2 = CA^2 + CB^2$. [The locus of the intersection of tangents at right angles is called the *Director Circle*.]

ELLIPSE.

PROPOSITION XVI. (*Aliter.*)

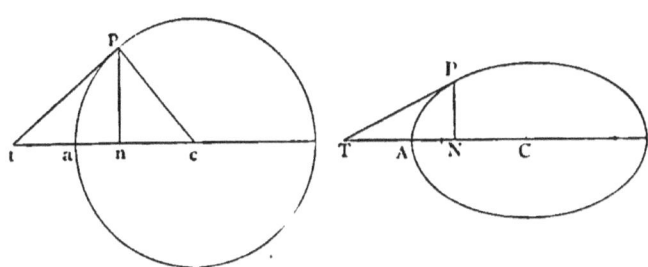

Draw the circle from which the ellipse is projected, and let C, P, T, N, A be the projections of

$$c, p, t, n, a.$$

Then pt touches the circle;   [Prop. δ.

therefore cpt is a right angle,   [Euc. III. 18.

and cnp is a right angle;   [Prop. ζ.

$\therefore cn \cdot ct = cp^2$;   [Euc. VI. 8.

$\therefore cn \cdot ct = ca^2$;

$\therefore CN \cdot CT = CA^2$.   [Prop. β.

Prop. XVI.

1. p is the point on the auxiliary circle corresponding to P. Sy is drawn perpendicular to the tangent at p. Prove $Sy = SP$.

2. Any circle through N, T, cuts the auxiliary circle at right angles.

3. If CY, AZ be the perpendiculars from the centre and an extremity of the major axis on the tangent to the ellipse at any point P, shew that
$$CA \cdot AZ = CY \cdot AN.$$

Proposition XVII.

If the tangent at P *meets the minor axis produced in* t, *and* Pn *is the perpendicular from* P *on the minor axis*

$$Cn \cdot Ct = CB^2.$$

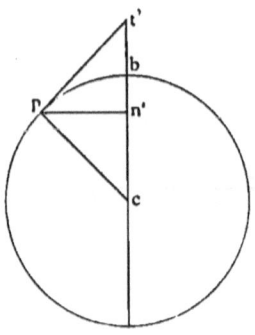
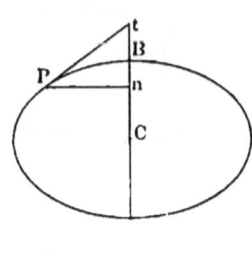

Draw the circle of which the ellipse is the projection.

And let c, p, t', b, n' be the points of which C, P, t, B, n are the projections.

Join cp. Then pt' touches the circle; [Prop. δ.

therefore cpt' is a right angle. [Euc. III. 18.

Also $cn'p$ is a right angle; [Prop ζ.

∴ $cn' \cdot ct' = cp^2$ [Euc. VI. 8.

$= cb^2$;

$Cn \cdot Ct = CB^2.$ [Prop. β.

ELLIPSE.

PROPOSITION XVIII.

If PF is the perpendicular from P on a line through C parallel to the tangent at P, and if the normal at P meets the minor axis in g, then

$$PF \cdot PG = CB^2 \text{ and } PF \cdot Pg = CA^2.$$

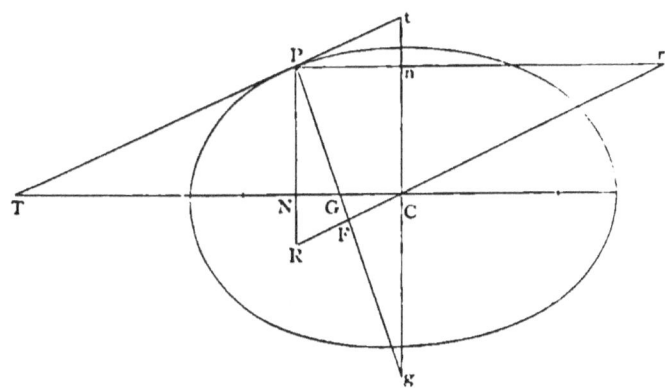

Draw PNR, Pnr perpendicular to the axes meeting CF in R and r, and let the tangent at P meet the axes at T and t.

Since the angles at N and F are right angles, a circle can be described through GNR and F; [Euc. III. 31.

$$\therefore PF \cdot PG = PN \cdot PR \quad \text{[Euc. III. 36.}$$
$$= Cn \cdot Ct \quad \text{[Euc. I. 34.}$$
$$= CB^2. \quad \text{[Prop. XVII.}$$

Similarly
$$PF \cdot Pg = Pn \cdot Pr$$
$$= CN \cdot CT \quad \text{[Euc. I. 34.}$$
$$= CA^2.$$

Prop. XVIII.

1. If from g a perpendicular gK be dropped on SP or $S'P$, prove that $PK = CA$.

2. If the tangent at P meets the major axis in T, then $CF \cdot PT$ is equal to the product of perpendiculars from the foci on the normal at P.

Proposition XIX.

$$GN : CN = CB^2 : CA^2.$$

Also $$CG = e^2 . CN.$$

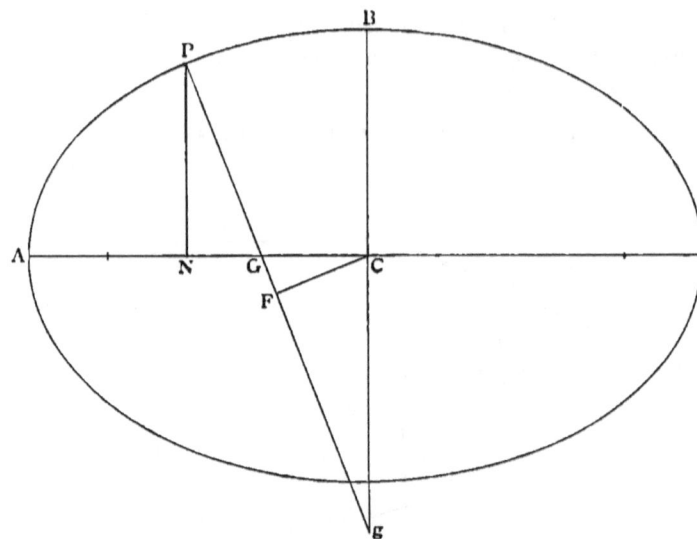

Produce PG to meet the minor axis at g, and draw CF parallel to the tangent at P, meeting Pg at F.

Then
$$GN : CN = PG : Pg \quad \text{[Euc. VI. 2.}$$
$$= PF.PG : PF.Pg$$
$$= CB^2 : CA^2. \quad \text{[Prop. XVIII.}$$

Also $CN - GN : CN = CA^2 - CB^2 : CA^2$;

$\therefore CG : CN = CS^2 : CA^2$; [Prop. IX.

$\therefore CG = e^2 . CN.$ [Prop. VII.

Prop. XIX.

1. If the tangent and normal at P meet the major and minor axes respectively in T, t, G, g, prove

 (a) $CG . CT = CS^2$,
 (b) $Cg . Ct = CS^2$,
 (c) Tg, tG are at right angles.

2. Prove $NG . CT = CB^2$.

3. From this proposition deduce the corresponding proposition for the parabola, viz. $NG = 2AS$.

4. Find a point P on the ellipse such that PG bisects the angle between CP and PN.

ELLIPSE. 57

PROPOSITION XX.

If from any point O on the tangent at P, OI is drawn perpendicular to the directrix, and OU perpendicular to SP, then $SU = e \cdot OI$. (Adams's property.)

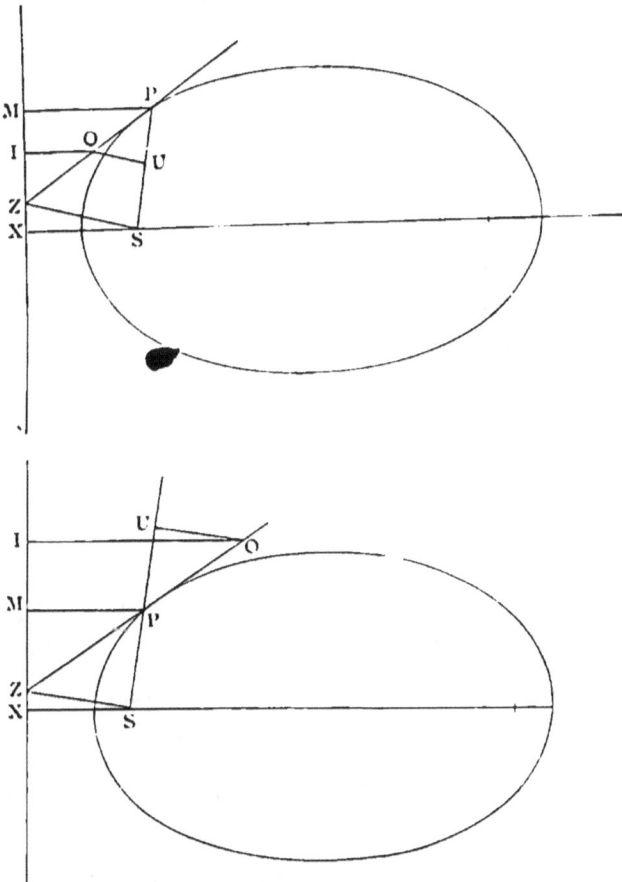

Join *SZ*, and draw *PM* perpendicular to the directrix.
ZSP is a right angle;   [Prop. XI.
∴ *ZS* is parallel to *OU*;
∴ $SU : SP = ZO : ZP$ [Euc. VI. 2.
            $= OI : PM$,   [Euc. VI. 4.
but $SP = e \cdot PM$;
∴ $SU = e \cdot OI$.

If the tangent at *P* meet the directrices in *Z*, *Z'*, the perpendiculars from *Z* and *Z'* on *SP* intercept a part equal to *AA'*.

Proposition XXI.

To draw a pair of tangents OQ, OQ' *to an ellipse from an external point* O.

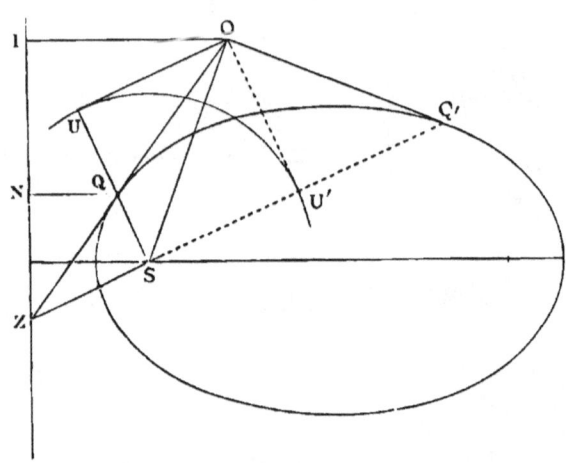

Draw OI perpendicular to the directrix.

With centre S, and radius $e \cdot OI$ describe a circle, and draw the tangents OU, OU'.　　　　　　　　[Euc. III. 17.

Draw SZ perpendicular to SU, meeting the directrix in Z. Join ZO, meeting SU in Q. Draw QN perpendicular to the directrix.　　　　　　　　　　　　　　[Euc. VI. 2.

Then
$$SQ : SU = QZ : OZ$$
$$= QN : OI ;$$
$$\therefore SQ : QN = SU : OI = e : 1;$$

therefore S is on the ellipse.

And since QSZ is a right angle, OQ touches the ellipse.
　　　　　　　　　　　　　　　　　　　　　[Prop. 11.

Similarly a second tangent OQ' may be drawn.

ELLIPSE.

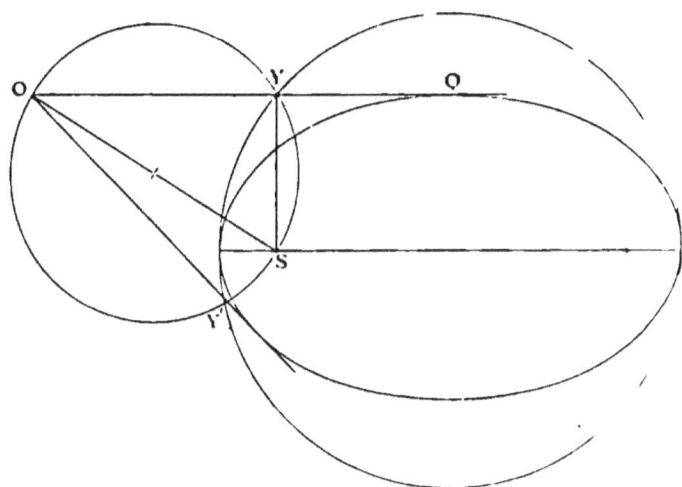

(*Second method.*) On OS as diameter describe a circle meeting the auxiliary circle in Y' and Y''. Then $SY'O$ is a right angle [Euc. III. 31], and OY' touches the ellipse [Prop. XIV.]. Similarly OY'' touches the ellipse.

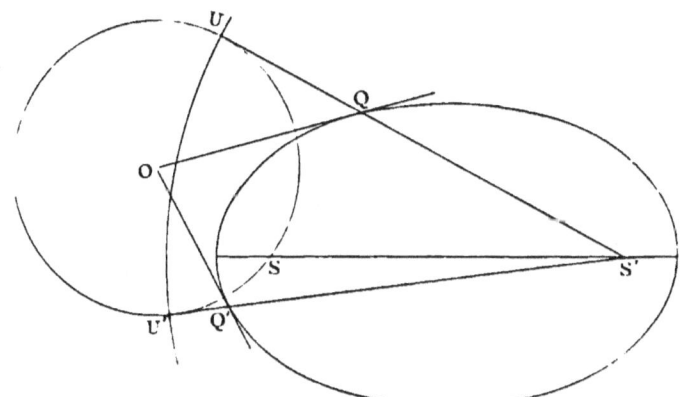

(*Third method.*) With centre O and radius OS describe a circle, and with centre S' and radius AA' describe a second circle intersecting the first in U and U'. Join $S'U$, $S'U''$ meeting the ellipse in Q and Q', then

$$\text{angle } OQU = \text{angle } OQS, \qquad [\text{Euc. I. 8.}$$
and OQ touches the ellipse. [Prop. XIII.

Similarly OQ' touches the ellipse.

Proposition XXII.

Tangents OQ, OQ' *subtend equal angles* OSQ, OSQ' *at the focus* S.

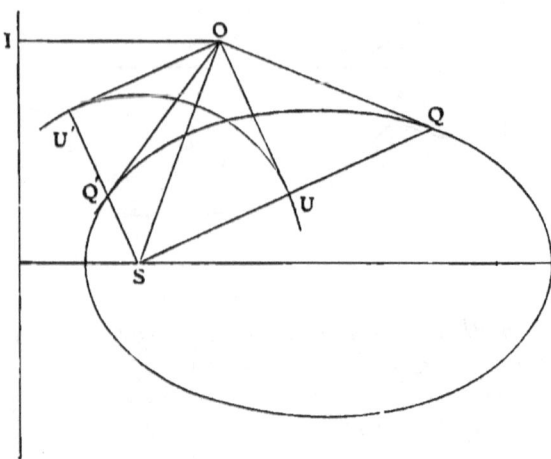

Draw OU, OU', OI perpendicular to SQ, SQ', and the directrix. Join OS.

Then
$$SU = e \cdot OI \qquad \text{[Prop. xx.}$$
$$= SU'; \qquad \text{[Prop. xx.}$$
$$\therefore\ OU = OU'; \qquad \text{[Euc. i. 47.}$$
$$\therefore\ OSU = OSU', \qquad \text{[Euc. i. 8.}$$
or
$$OSQ = OSQ'.$$

Prop. XXII.

1. QQ' produced meets the directrix in K, prove that OSK is a right angle.

2. Tangents at the extremities of a focal chord meet the tangent at the vertex in T_1, T_2, prove $AT_1 \cdot AT_2 = AS^2$.

3. OQ, OQ' are two fixed tangents to an ellipse. A variable tangent intersects them in q, q'. Prove that the angle qSq' is constant.

4. Normals at the extremities of a focal chord meet in W, and the corresponding tangents in Z. Prove that ZW passes through the other focus.

5. OQ, OQ' are tangents from O, and OS meets QQ' in R. RZ, parallel to the axis, meets the directrix in Z. Shew that QZ and $Q'Z$ are equally inclined to the axis.

Proposition XXIII.

Tangents OQ, OQ′ *are inclined at equal angles to* OS, OS′.

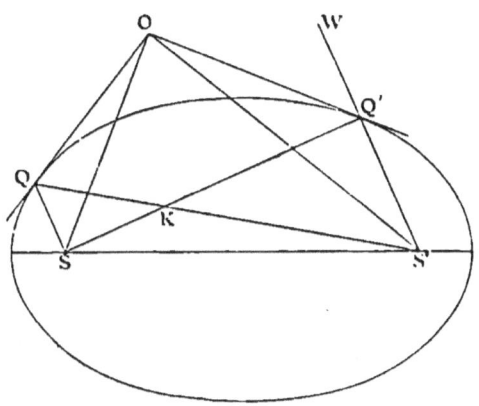

Join SQ, SQ', $S'Q$, $S'Q'$ and produce $S'Q'$ to W, and let SQ' meet $S'Q$ in K.

Then angle $S'OQ' = OQ'W - OS'Q'$ [Euc. I. 32.
$\qquad\qquad = \tfrac{1}{2}SQ'W - \tfrac{1}{2}QS'Q'$ [Props. XIII., XXII.
$\qquad\qquad = \tfrac{1}{2}S'KQ'$.   [Euc. I. 32.

Similarly $\quad SOQ = \tfrac{1}{2}SKQ$;

$\therefore\quad SOQ = S'OQ'$.   [Euc. I. 15.

Prop. XXIII.

1. Given two tangents to an ellipse and one focus, find the locus of the centre.

2. On OQ, OQ', lengths OR, OR' are taken, equal to OS, OS' respectively. Prove that RR' is equal to the major axis of the ellipse.

62   ELLIPSE.

Proposition XXIV.

The locus of the middle points of any system of parallel chords of an ellipse is a straight line passing through the centre; and the tangent at either end of the straight line is parallel to the chords.

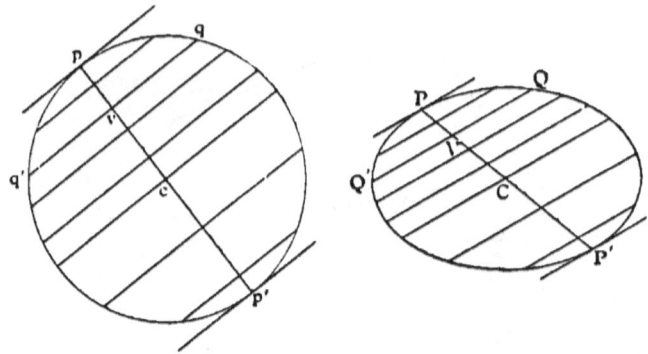

Draw the circle whose projection is the ellipse. The middle points of the system of parallel chords of the ellipse are the projections of the middle points of a system of parallel chords of the circle.   [Props. β and γ.

In the circle these middle points lie on a straight line cv passing through the centre c.   [Euc. III. 3.

And the projection of cv is a straight line CV passing through the centre C of the ellipse.   [Prop. a.

In the circle the tangents at either end of cv are parallel to the chords, because they are all perpendicular to cv.
         [Euc. III. 3 and 16.

Hence in the ellipse the same is true.   [Props. γ and δ.

Def. The locus of the middle point of a system of parallel chords is called a *diameter*.

Note. The words diameter and axis are frequently used to denote the length of the portion of the diameter or axis intercepted by the curve.

Def. The half (QV) of a chord (QQ') which is bisected by a diameter (CP) is called an *ordinate to the diameter*.

Proposition XXV.

Tangents at the ends of any chord meet on the diameter which bisects the chord.

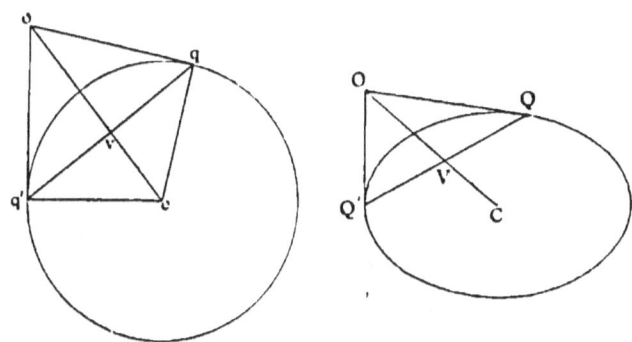

Let OQ, OQ' be the tangents, join CO, meeting QQ' in V.

Draw the circle whose projection is the ellipse, and let o, Q, Q', C, V be the projections of o, q, q', c, v. Join cq, cq'.

Then oq, oq' touch the circle;   [Prop. 8.

$\therefore oq = oq'$;   [Euc. III. 36.

$\therefore$ angle $ocq =$ angle ocq';   [Euc. I. 8.

$\therefore qv = q'v$;   [Euc. I. 4.

$\therefore QV = Q'V$.   [Prop. 3.

Prop. XXV.

1. The tangent at a point P of an ellipse meets the tangent at A in Y. Shew that CY is parallel to $A'P$.

2. If CP meets the directrix in Z, ZS is perpendicular to QQ'.

Proposition XXVI.

QV *is an ordinate of the diameter* CP; *if the tangent at* Q *meets the diameter* CP *produced in* O, *then*
$$CV \cdot CO = CP^2.$$

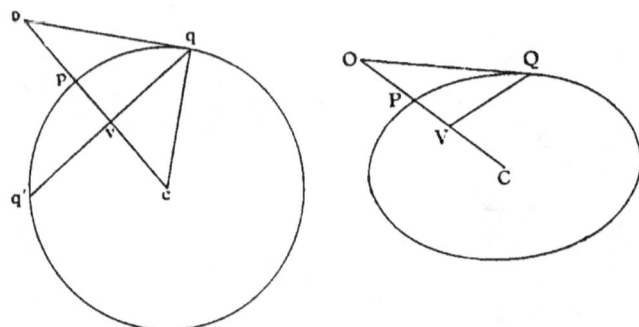

Draw the circle whose projection is the ellipse. Let c, q, o, p, v be the projections of C, Q, O, P, V. Join cq and produce qv to meet the circle at q'.

Then oq is a tangent,   [Prop. δ.

  qq' is bisected at v,   [Prop. β.

$\therefore$ cvq is a right angle,   [Euc. III. 3.

and cqo is a right angle,   [Euc. III. 18.

$\therefore$ $cv \cdot co = cq^2$,   [Euc. VI. 8.

$\therefore$ $cv \cdot co = cp^2$,

$\therefore$ $CV \cdot CO = CP^2$.   [Prop. β.

Prop. XXVI.

1. PR parallel to PQ meets CQ in R. Prove that PR is parallel to the tangent at Q.

2. The tangent at any point P of an ellipse meets the equiconjugate diameters [see page 66] in T and T'. Shew that the triangles TCP, $T'CP$ are in the ratio $CT^2 : CT'^2$.

Proposition XXVI. (*Aliter.*)

Draw the tangent at P meeting QO in R.
Draw PW parallel to OQ meeting QV in W.
Join PQ, RW.

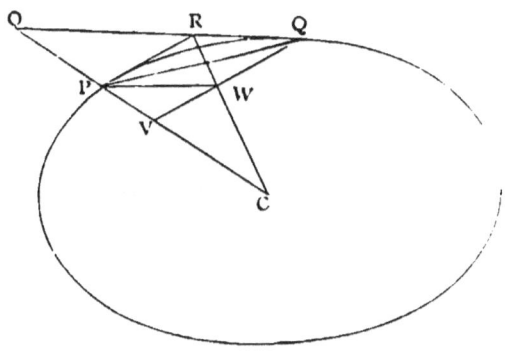

Then ∵ $PRQW$ is a parallelogram,

∴ RW bisects PQ,

∴ RW passes through the centre,   [Prop. 25.

∴ by similar triangles

$$CV : CP = CW : CR$$
$$= CP : CO,$$
∴ $CV \cdot CO = CP^2.$

What is the corresponding proposition in the parabola? Apply this method of proof to it.

This proof is due to the Master of St John's College, Cambridge.

Proposition XXVII.

If CP *bisects chords parallel to* CD, *then* CD *bisects chords parallel to* CP.

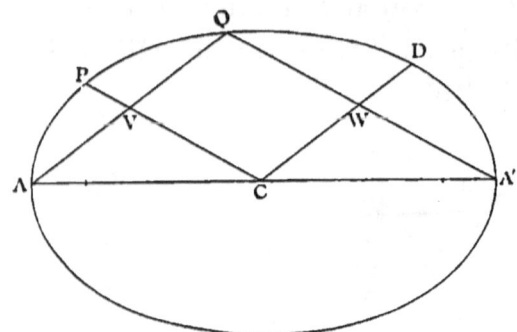

Draw AQ parallel to CD meeting CP in V;
    then AQ is bisected at V.
Join $A'Q$ cutting CD in W.
Since AQ is bisected in V
    and AA' in C,
  ∴ $A'Q$ is parallel to CP.
And ∵ CD is parallel to AQ,
    and AA' is bisected in C,
  ∴ $A'Q$ is bisected in W,
∴ CD bisects the chord $A'Q$ which is parallel to CP,
∴ CD bisects all chords parallel to CP.   [Prop. 24.

Def. Two diameters which are so related that each bisects chords parallel to the other are called *conjugate diameters*.

N.B. The tangent at P is parallel to CD and the tangent at D is parallel to CP.   [Prop. 24.

Prop. XXVII.

1. To draw the equiconjugate diameters of the ellipse.
2. The focus is the centre of perpendiculars of the triangle formed by two conjugate diameters and the directrix.

ELLIPSE.

Proposition XXVIII.

Conjugate diameters in the ellipse are the projections of diameters in the circle at right angles to one another.

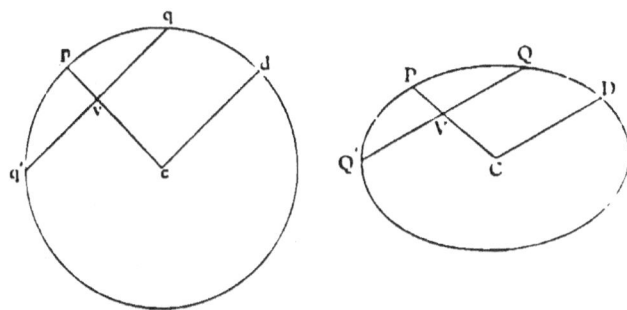

Let CP, CD be conjugate diameters. Draw a chord QVQ' parallel to CD and bisected at V. Draw the circle whose projection is the ellipse and let D, Q, P, Q', V, C be the projections of d, q, p, q', v, c.

cd is parallel to qq',   [Prop. γ.

and qq' is bisected at v,   [Prop. β.

∴ cv is perpendicular to qq',   [Euc. III. 3.

∴ cp is perpendicular to cd.

NOTE. Numerous metrical properties of conjugate diameters may be deduced from this proposition by the method used in Prop. xxx., e.g.:

1. $P'CP$, CD are two conjugate diameters, R any other point on the ellipse. PR, $P'R$ meet CD or CD produced in T, t. Prove $CT . Ct = CD^2$.

2. If CP, CD, CQ, CR be two pairs of conjugate diameters, and if the tangent at P meet CQ, CR produced in T, t; then $PT . Pt = CD^2$.

5—2

ELLIPSE.

DEF. Chords (QP, QP'), which join any point (Q) on an ellipse to the extremities of a diameter (PCP') are called *supplemental chords*.

PROPOSITION XXIX.

Supplemental chords are parallel to conjugate diameters.

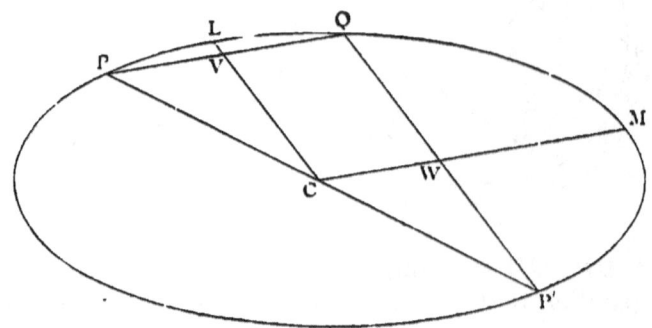

Draw the diameters CL, CM parallel to the supplemental chords $P'Q$, QP cutting them in V and W.

Then $PV : VQ = PC : CP'$,   [Euc. VI. 2.

∴ $PV = VQ$,

∴ CL bisects all chords parallel to PQ,   [Prop. 24.

that is parallel to CM.

Similarly CM bisects all chords parallel to CL.

∴ CL, CM are conjugate diameters.

The diagonals of any parallelogram circumscribed to an ellipse are conjugate diameters.

ELLIPSE.

Proposition XXX.

QV *is an ordinate of the diameter* PCP', CD *the diameter parallel to* QV, *then*

$$QV^2 : PV \cdot P'V = CD^2 : CP^2.$$

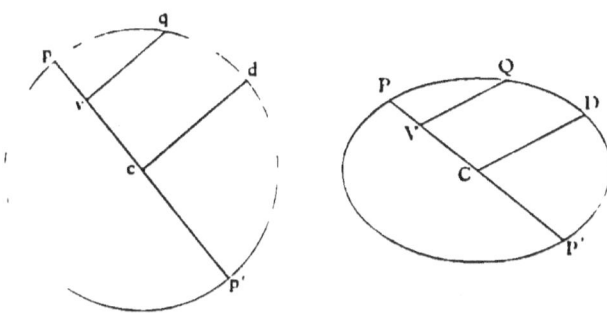

Draw the circle whose projection is the ellipse, and let P, V, C, P', Q, D be the projections of p, v, c, p', q, d.

Since CP, CD are conjugate diameters

$\qquad\qquad pcd$ is a right angle. [Prop. 28.

But $\qquad qv$ is parallel to cd. [Prop. γ.

Hence $\qquad qv$ is perpendicular to cp,

$\qquad\qquad \therefore qv^2 = pv \cdot p'v,$ [Euc. III. 3 and 35.

$\qquad\qquad \therefore qv^2 : pv \cdot p'v = cd^2 : cp^2,$

but $\qquad qv^2 : cd^2 = QV^2 : CD^2,$ [Prop. γ.

$\qquad\qquad pv \cdot p'v : cp^2 = PV \cdot P'V : CP^2,$ [Prop. γ.

$\qquad\qquad \therefore QV^2 : PV \cdot P'V = CD^2 : CP^2.$

On QV or QV produced is taken a point R, such that $VR : VQ = CP : CD$. Shew that the locus of R is an ellipse, and find the position of its axes.

ELLIPSE.

PROPOSITION XXXI.

In the triangles CPN, CDR, CR : PN = CA : CB
and CN : DR = CA : CB.

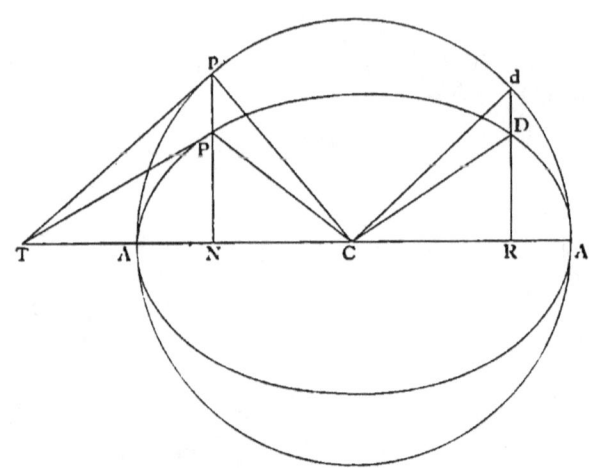

Draw the auxiliary circle.
Produce NP, RD to meet it in p and d.
Join Cp, Cd and draw the tangents pT, PT to the circle and ellipse respectively, intersecting on the axis. [Prop. 15.
Then PT is parallel to CD, [Prop. 24.
 ∴ the triangles TNP, CRD are similar,
 ∴ $TN : CR = NP : RD = Np : Rd$, [Prop. 4.
and the angle TNp = the angle CRd,
 ∴ triangles TNp, CRd are similar, [Euc. VI. 6.
 ∴ pT is parallel to Cd,
 ∴ the angle pCd = angle CpT = a right angle,
therefore the angles NpC, dCR are equal, each being the complement of angle pCN,
 ∴ the triangles pNC, CRd are equal in all respects, [Euc. I. 26.
 ∴ $pN = CR$.
But $pN : PN = CA : CB$,
 ∴ $CR : PN = CA : CB$.
Similarly $CN : DR = CA : CB$.

ELLIPSE.

PROPOSITION XXXII.

$$CP^2 + CD^2 = CA^2 + CB^2.$$

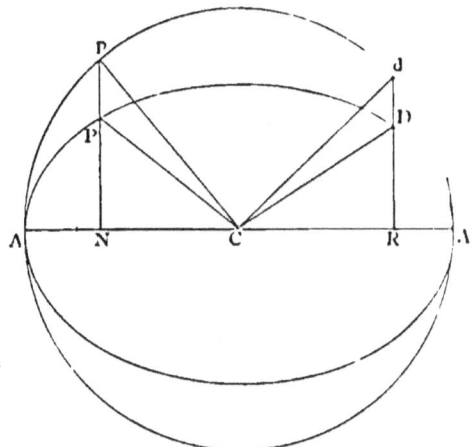

Draw the auxiliary circle.
Produce NP, RD to meet it in p and d.
Join Cp, Cd.
Then $\qquad DR^2 : CN^2 = CB^2 : CA^2,$ [Prop. 31.
and $\qquad PN^2 : CR^2 = CB^2 : CA^2,$ [Prop. 31.
$\qquad \therefore DR^2 + PN^2 : CN^2 + CR^2 = CB^2 : CA^2.$
But $\qquad CN^2 + CR^2 = CN^2 + pN^2 = CA^2,$ [Prop. 31.
$\qquad \therefore DR^2 + PN^2 = CB^2.$
Now $\qquad CP^2 + CD^2 = CR^2 + CN^2 + DR^2 + PN^2$
$\qquad\qquad\qquad = CA^2 + CB^2.$

PROP. XXXI.

If the tangent at P meet the major axis in T, and if Q be the foot of the perpendicular from C on the tangent, prove that
$$CQ \cdot QT : CT^2 = CN \cdot PN : CD^2.$$
Prove
(a) $PG : CD = CB : CA$;
(b) $Pg : CD = CA : CB$;
(c) $PG \cdot Pg = CD^2.$

PROP. XXXII.

1. Find the greatest and least values of the sum of a pair of conjugate diameters.

2. CP, CD are conjugate diameters. If PG, DH be the normals at P and D, prove that $PG^2 + DH^2$ is constant.

Proposition XXXIII.

The area of the parallelogram formed by tangents at the extremities of a pair of conjugate diameters is constant.

$$PF \cdot CD = CA \cdot CB.$$

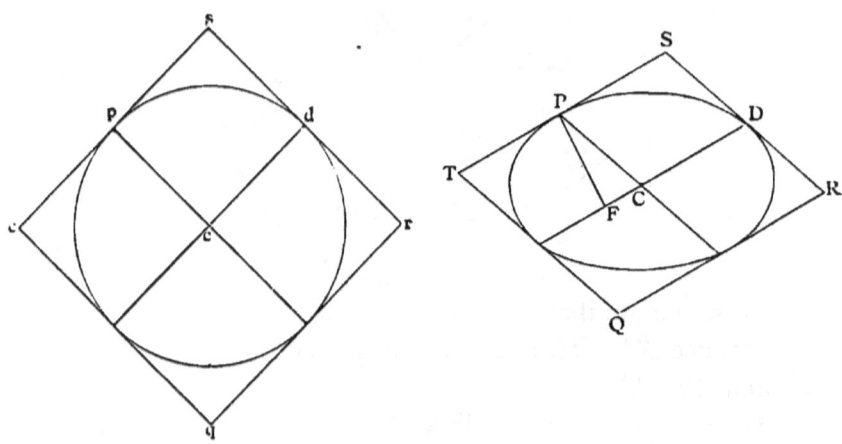

Let $QRST$ be the circumscribing parallelogram, then its sides are parallel to CP or CD. [Prop. 24.

Draw the circle, whose projection is the ellipse, and let p, c, d, q, r, &c. be the points whose projections are P, C, D, Q, R, &c.

Then pcd is a right angle, because CP, CD are conjugate to one another, [Prop. 28.

$qrst$ circumscribes the circle, [Prop. δ.

and its sides are parallel to cp or cd, [Prop. γ.

hence $qrst$ is a square, equal to the square on the diameter and constant in area.

Hence $QRST$ is also constant. [Prop. ε.

Again this parallelogram is equal to $4PF \cdot CD$, but if CP, CD are the axes, the area is $4CA \cdot CB$,

$$\therefore PF \cdot CD = CA \cdot CB.$$

ELLIPSE.

PROPOSITION XXXIV.

If two chords of an ellipse intersect, the rectangles contained by their segments are as the squares of the parallel semi-diameters.

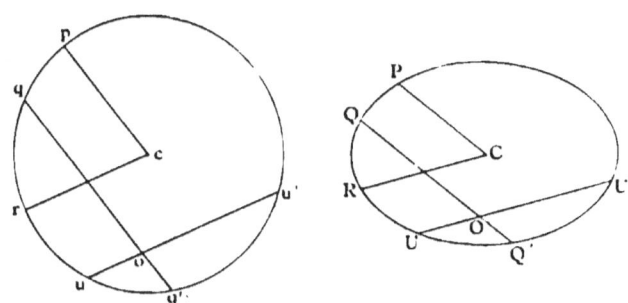

Let QOQ', UOU' be the chords and CP, CR the parallel semi-diameters.

Draw the circle whose projection is the ellipse, and let q, o, q', &c. be the points whose projections are Q, O, Q', &c.

In the circle $qo \cdot oq' = uo \cdot ou'$, [Euc. III. 35.

and $\qquad\qquad cp^2 = cr^2$,

$\therefore qo \cdot oq' : uo \cdot ou' = cp^2 : cr^2$,

but $\qquad qo \cdot oq' : cp^2 = QO \cdot OQ' : CP^2$, [Prop. γ.

and $\qquad uo \cdot ou' : cr^2 = UO \cdot OU' : CR^2$, [Prop. γ.

$\therefore QO \cdot OQ' : UO \cdot OU' = CP^2 : CR^2$.

Prop. XXXIII.

1. $PG \cdot Pg = CD^2$. (See Prop. 18.)
2. $SP \cdot S'P = CD^2$.
3. $CD \cdot SY = BC \cdot SP$.
4. CD is conjugate to CP. If DQ be drawn parallel to SP, and CQ perpendicular to DQ, prove that CQ is equal to the semi-axis minor.
5. From D tangents are drawn to the circle on the minor axis as diameter. Prove that these tangents are parallel to the focal distances of P.

Prop. XXXIV.

1. The tangents to an ellipse from an external point are proportional to the parallel semi-diameters.

2. If a circle intersect an ellipse in four points, the chords of intersection are equally inclined to the axis.

3. If a circle touch an ellipse at the points P and Q, shew that PQ is parallel to one of the axes.

4. Deduce Prop. 3 and Prop. 30 from Prop. 34.

5. If PQ, PQ' are chords equally inclined to the axis, prove that the circle circumscribing PQQ' touches the conic at P.

HYPERBOLA.

Def. A *hyperbola* is the locus of a point (P) whose distance from a fixed point (S) bears a constant ratio (e), greater than unity, to its distance (PM) from a fixed straight line (XM),

$$(SP = e \cdot PM).$$

The fixed point (S) is called the *focus*.

The fixed straight line (XM) is called the *directrix*.

The constant ratio (e) is called the *eccentricity*.

HYPERBOLA.

PROPOSITION I.

Construction for points on the hyperbola.
The perpendicular on the directrix through the focus is an axis of symmetry.
To find the vertices A and A'.

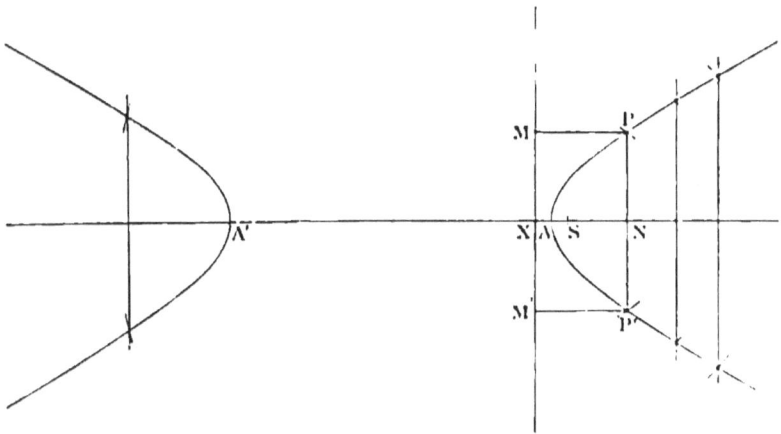

From the focus S draw SX perpendicular to the directrix. Divide XS in A, so that
$$SA = e \cdot AX;$$
also in SX produced take A' so that
$$SA' = e \cdot A'X.$$
Then A and A' are points on the curve.

Take any point N on the straight line AA', with centre S and radius $e \cdot NX$ describe a circle, through N draw PNP' perpendicular to AA' and cutting the circle in P and P', then P and P' are points on the hyperbola. Draw PM, $P'M'$ perpendicular to the directrix,
$$SP = e \cdot NX = e \cdot PM,$$
$$SP' = e \cdot NX = e \cdot P'M'.$$

Corresponding to any point N on the line AA', we thus get two points P and P' at equal distances on opposite sides of AA'; hence the hyperbola is symmetrical with respect to AA', or AA' is an axis, and the points A and A' are vertices.

NOTE. It may be proved that the circle intersects the perpendicular NP, when N is in any part of the axis AA', except the part between A and A', hence the hyperbola lies entirely outside the lines through A and A' perpendicular to the axis, but it is infinitely extended in both directions (see Appendix).

HYPERBOLA.

Proposition II.

If the chord PP′ *intersects the directrix in* K, SK *bisects the angle between* SP *and* SP′.

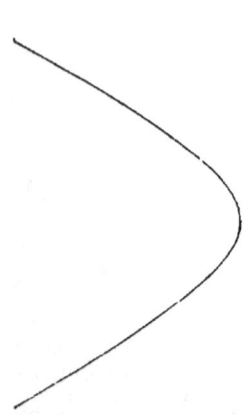
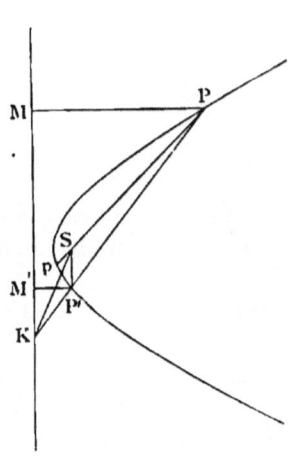

Join SP, SP', SK; produce PS to p, and draw $PM, P'M'$ perpendicular to the directrix.

Then $\qquad SP = e \cdot PM,$
and $\qquad SP' = e \cdot P'M'$;
$\therefore SP : SP' = PM : P'M'$
$\qquad\qquad = PK : P'K,$

by similar triangles $PKM, P'KM'$.

Therefore SK bisects $P'Sp$. (Euc. VI. A.)

HYPERBOLA.

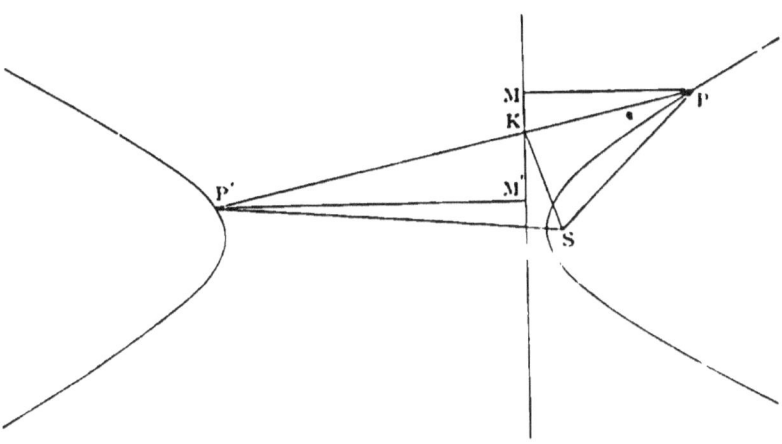

Similarly if P and P' are on opposite branches of the hyperbola SK bisects the angle PSP'.

Prove that a st. line cuts the hyperbola in two points only.

Prop. I.

1. In any conic, if PR be drawn to the directrix parallel to a fixed straight line, the ratio $SP : PR$ is constant.

2. If an ellipse, a parabola, and a hyperbola have the same focus and directrix, the ellipse will be entirely on one side of the parabola, and the hyperbola on the other.

3. In any conic a chord through the focus is divided harmonically by the focus and directrix.

Prop. II.

1. Prove that a straight line can cut a conic in two points only.

2. In any conic if two fixed points PP'' on the curve be joined to a variable point Q, and PQ, $P''Q$ meet the directrix in p, p', the angle pSp' is constant.

Proposition III.

If PN *is the ordinate of a point* P *on the hyperbola*,

$$PN^2 : AN . A'N$$

is a constant ratio.

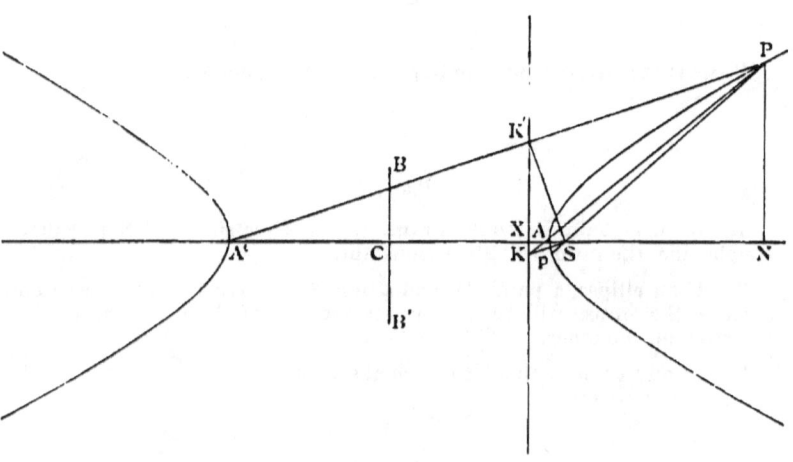

Join PA, $A'P$, and let them, produced if necessary, meet the directrix at K and K'.

Join SP, SK, SK', and produce PS to p.

By similar triangles PAN, KAN,

$$PN : AN = KN : AN.$$

By similar triangles $PA'N$, $K'A'N$,

$$PN : A'N = K'N : A'N;$$

$$\therefore PN^2 : AN.A'N = KN.K'N : AN.A'N.$$

But SK bisects the angle ASp, [Prop. 2.
and SK' bisects the angle ASP, [Prop. 2.

$\therefore KSK'$ is a right angle;

$$\therefore KN.K'N = SN^2; \quad \text{[Euc. vi. 8.}$$

$$\therefore PN^2 : AN.A'N = SN^2 : AN.A'N,$$

which is a constant ratio.

Def. Take $CB^2 : CA^2$ in this constant ratio, drawing CB perpendicular to AA'.

I. Then AA' is called the *transverse axis*.

II. C is called the *centre* of the curve.

III. CB is called the *semi-conjugate axis*.

So that $PN^2 : AN.A'N = CB^2 : CA^2$.

Prop. III.

1. PNP' is a double ordinate of an ellipse. Find the locus of the intersection of AP and $A'P'$.

2. In the rectangular hyperbola (page 81) $PN^2 = AN.A'N$.

3. PNP' is a double ordinate of a rectangular hyperbola. Prove the angles PAP', $PA'P'$ are supplementary.

4. The tangent at any point P of a circle meets a fixed diameter AB produced in T. Shew that the straight line through T perpendicular to this diameter will cut AP, BP produced in points which lie upon a certain rectangular hyperbola.

HYPERBOLA.

Proposition IV.

If the diagonals of the rectangle, formed by perpendiculars through the extremities of the axes ACA′, BCB′, *be produced indefinitely, and the ordinate* NP *be produced both ways to meet them in* p, p′, *the rectangle* Pp . Pp′ = CB².

Also the curve continually approaches to each diagonal without actually meeting it, and its distance from it becomes ultimately less than any finite length.

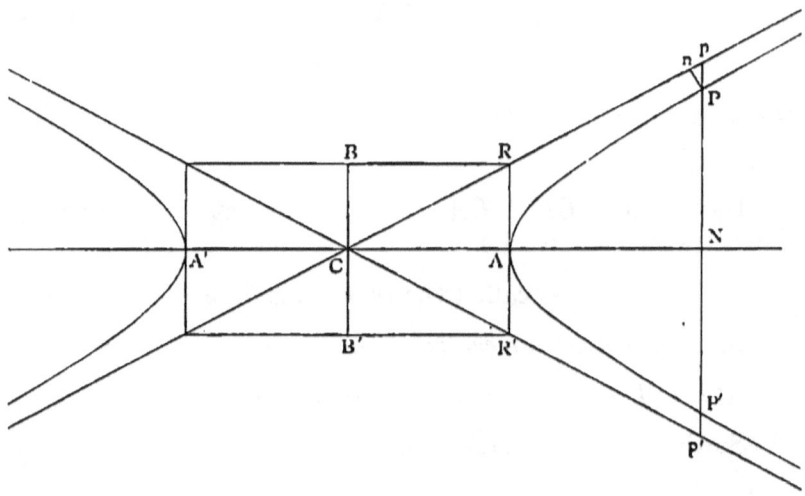

Let parallels to the axes through A and B meet in R, and let Pp' meet the curve at P'.

Then PP', pp' are both bisected in N; [Prop. 1.

$$\therefore pP' = p'P.$$

But $\quad pP \cdot pP' = Np^2 - NP^2;$ [Euc. II. 5.

$$\therefore pP \cdot p'P = Np^2 - NP^2.$$

Now $pN^2 : CN^2 = AR^2 : CA^2$
$= CB^2 : CA^2$.

Again $PN^2 : AN . A'N = CB^2 : CA^2$, [Prop. 3.

or $PN^2 : CN^2 - CA^2 = CB^2 : CA^2$. [Euc. II. 6.

Subtracting $pN^2 - PN^2 : CA^2 = CB^2 : CA^2$;

$\therefore pN^2 - PN^2 = CB^2$;

$\therefore pP . p'P = CB^2$.

Since the product $pP . p'P$ is constant, of which one factor $p'P$ constantly increases therefore pP constantly diminishes and finally becomes less than any finite quantity. And if Pn be drawn perpendicular to CR the ratio $Pn : Pp$ is constant, therefore Pn continually diminishes and finally becomes less than any finite length.

DEF. When a curve continually approaches to a fixed straight line without ever actually meeting it, but so that its distance from it becomes ultimately less than any finite length, the line is said to be a *rectilinear asymptote* to the curve.

DEF. When the asymptotes of a hyperbola are at right angles the curve is called the *Rectangular Hyperbola*. In the Rectangular Hyperbola the axes are evidently equal. Hence the curve is sometimes called the *Equilateral Hyperbola*.

(NOTE. We shall use the abbreviation R. H. for Rectangular hyperbola.)

PROP. IV.

The circle on AA' as diameter cuts the directrices in the same points as the asymptotes.

Proposition V.

The curve is symmetrical with respect to the conjugate axis, and has a second focus and directrix.

Also all chords passing through C are bisected at C.

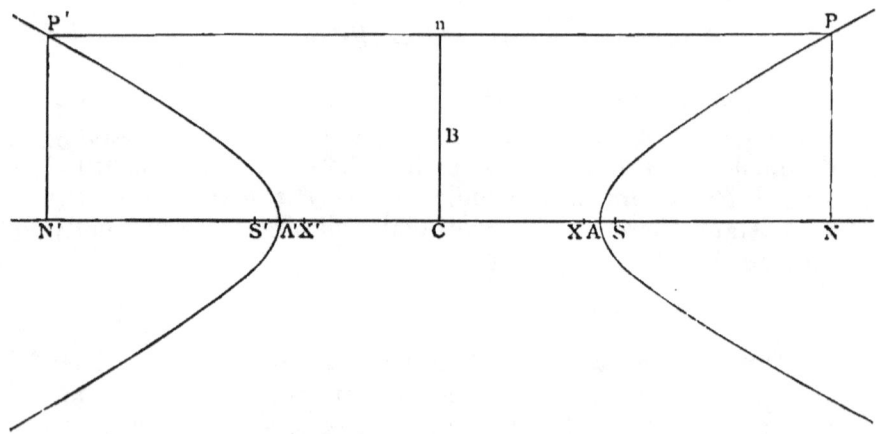

Draw the ordinate PN and take $CN' = CN$.

Since P is on the hyperbola, CN is $> CA$;

$$\therefore CN' \text{ is } > CA';$$

therefore a perpendicular through N' will cut the hyperbola. Let it cut it in P'.

Then $P'N'^2 : AN'.A'N' = PN^2 : AN.A'N$. [Prop. 3.

But $\qquad A'N' = AN$ and $AN' = A'N$;

$$\therefore AN'.A'N' = AN.A'N;$$

$$\therefore P'N'^2 = PN^2;$$

$$\therefore P'N' = PN.$$

Join PP', cutting CB or CB produced in n.

Therefore $P'nP$ is parallel to the axis, and therefore perpendicular to BC, and $Pn = P'n$.

Hence corresponding to any P on the hyperbola, there is another point P' on the hyperbola on the opposite side of CB, such that PP' is bisected at right angles by CB, or the hyperbola is symmetrical with respect to the conjugate axis.

If we take CS' equal to CS, and CX' equal to CX, and through X' draw a line perpendicular to AA', the hyperbola can be described with this line as directrix, S' as focus, and eccentricity the same as before.

PROP. VI. (See page 84.)

1. If an asymptote meets the directrix in E, $CE = CA$, and CES is a right angle.

2. If Pp be drawn parallel to an asymptote to meet the directrix in p, $Pp = SP$.

3. Having given the transverse and conjugate axis, find the focus and directrix.

Proposition VI.

$SA = e \cdot AX$; $CA = e \cdot CX$; $CS = e \cdot CA$; $CA^2 = CS \cdot CX$.

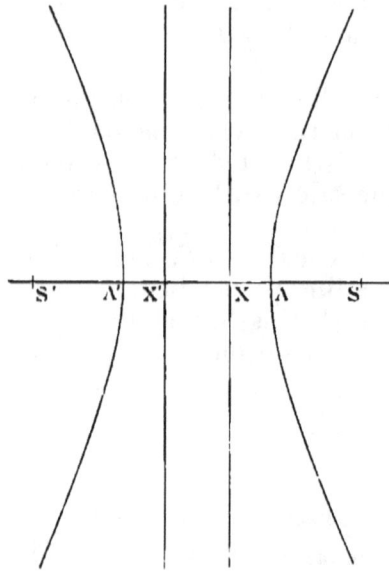

Because A and A' are points on the hyperbola;

$$\therefore SA = e \cdot AX, \qquad \text{[Def.}$$
$$SA' = e \cdot A'X \qquad \text{[Def.}$$
$$= e \cdot AX'.$$

By subtraction, $AA' = e \cdot XX'$,
$$\therefore CA = e \cdot CX \quad \ldots\ldots\ldots\ldots(\alpha).$$

By addition, $SS' = e \cdot AA'$,
$$\therefore CS = e \cdot CA \quad \ldots\ldots\ldots\ldots(\beta).$$
$$\therefore CA^2 = CS \cdot CX \quad \ldots\ldots\ldots\ldots(\gamma).$$

Note. In this figure the eccentricity is about 2·2, in the figure of prop. 5 the eccentricity is only 1·1, the student should observe the effect of this on the relative positions of S, A, X, and on the general shape of the curve. In this figure $CB = 2 \cdot CA$; in the figure of the last proposition $CA = 2 \cdot CB$.

Proposition VII.

$S'P \sim SP = AA'$. *Mechanical construction for hyperbola.*

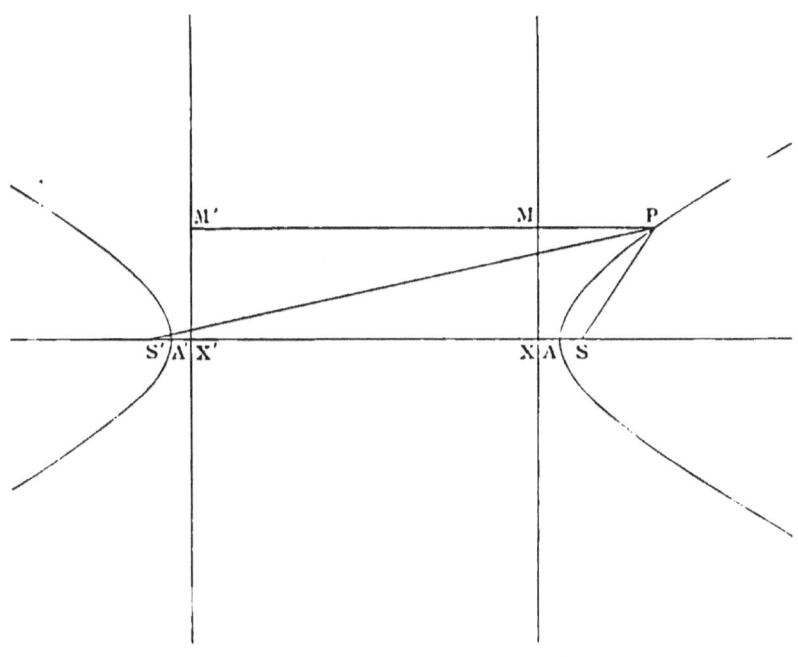

Draw PMM' perpendicular on the directrices.

Then
$$SP = e \cdot PM,$$
and
$$S'P = e \cdot PM';$$
$$\therefore S'P \sim SP = e \cdot MM'$$
$$= e \cdot XX'$$
$$= AA'.$$

Proposition VII. (*continued*).

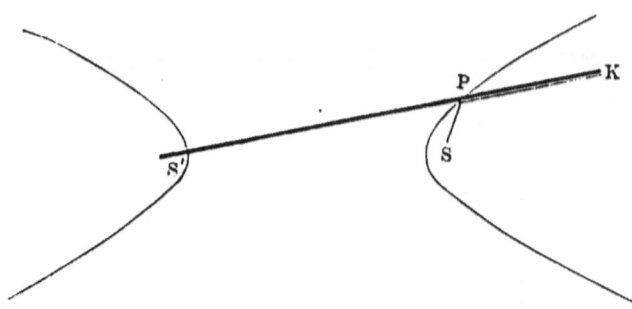

Hence the mechanical construction,

$S'K$ is a bar of wood hinged at S', and SPK a string stretched tight at P and fastened at S and K.

$$S'P + PK = \text{constant},$$

also
$$SP + PK = \text{constant},$$

$$\therefore S'P - SP = \text{constant}.$$

Prop. VII.

1. The locus of the centre of a circle which touches two fixed circles is an ellipse or hyperbola.

2. Given one focus of an ellipse and two points on the curve, the locus of the other focus is an hyperbola.

NOTE. The figures of this chapter have been drawn by using a wooden cone cut by a plane perpendicular to the base. See prop. 3 of the next chapter.

Proposition VIII.

$$CB^2 = CS^2 - CA^2 = SA \cdot SA'.$$

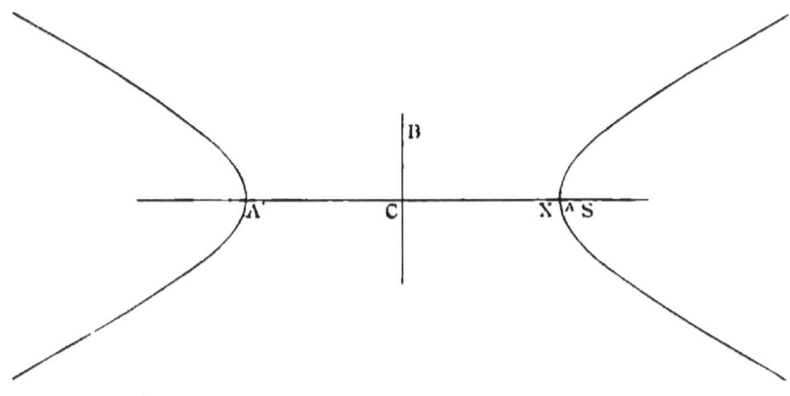

$$CS : CA = SA : AX, \qquad \text{[Prop. 6.}$$
$$\therefore CS + CA : CA = SA + AX : AX$$
$$= SX : AX \quad \ldots\ldots\ldots\ldots\ldots (1).$$
$$CS : CA = SA' : A'X, \qquad \text{[Prop. 6.}$$
$$\therefore CS - CA : CA = SA' - A'X : A'X$$
$$= SX : A'X \quad \ldots\ldots\ldots\ldots\ldots (2).$$

Therefore, multiplying (1) and (2) together,
$$CS^2 - CA^2 : CA^2 = SX^2 : AX \cdot A'X$$
$$= CB^2 : CA^2; \qquad \text{[Prop. 3.}$$
$$\therefore CS^2 - CA^2 = CB^2 = AS \cdot A'S. \qquad \text{[Euc. II. 5.}$$

Prop. VIII.

1. In the r. h. $e = \sqrt{2}$, $CS^2 = 2AC^2$ and $CS = 2CX$.

2. If the asymptote meet the directrix in E, and the tangent at the vertex in H, $SE = BC$, and SH is parallel to AE.

The *latus rectum* (*LL'*) is the double ordinate through the focus.

Proposition IX.

$$SL \cdot CA = CB^2.$$

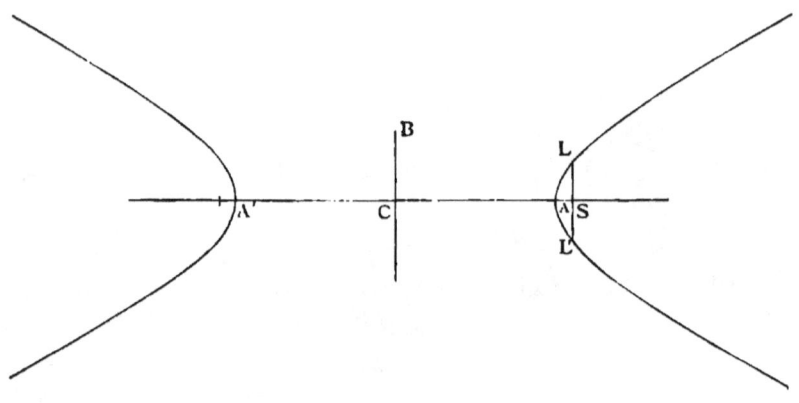

$$\begin{aligned}
SL^2 &: AS \cdot A'S = CB^2 : CA^2. \quad &[\text{Prop. 3.}] \\
\text{But} \quad AS \cdot A'S &= CB^2, \quad &[\text{Prop. 8.}] \\
\therefore SL^2 &: CB^2 = CB^2 : CA^2; \\
\therefore SL &: CB = CB : CA; \\
\therefore SL \cdot CA &= CB^2.
\end{aligned}$$

Prop. IX.

1. Prove this Prop. by means of props. 6 and 8.
2. In the r. h. $SL = CA$.

HYPERBOLA.

PROPOSITION X.

If the tangent at P *meets the directrix in* Z, PSZ *is a right angle.*

Also tangents at the ends of a focal chord intersect on the directrix.

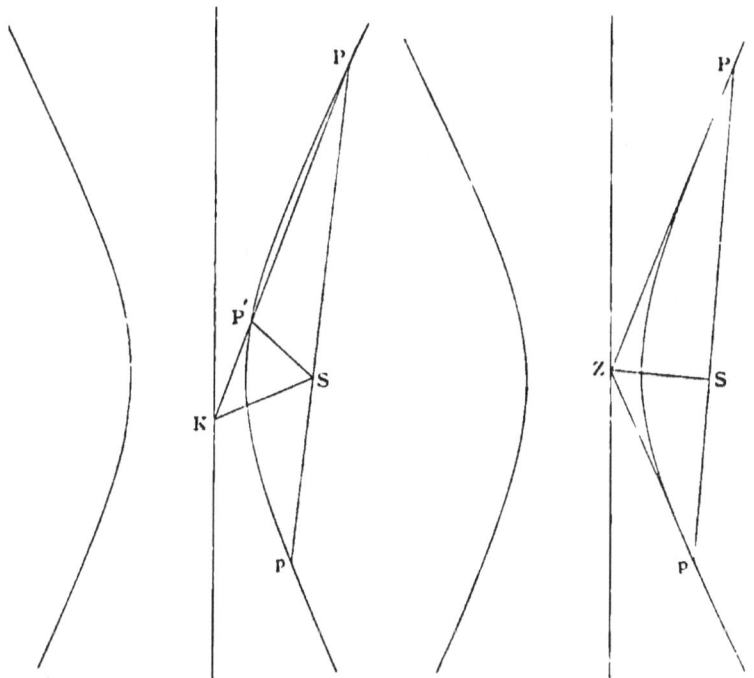

Take a point P' on the hyperbola near to P, and let the chord PP' meet the directrix in K, and produce PS to p. Then KS bisects the angle $P'Sp$. [Prop. 2.

When P' coincides with P (as in figure 2), so that $PP'K$ becomes the tangent PZ, and SK coincides with SZ, $P'Sp$ becomes two right angles; and PSZ is a right angle.

Hence ZSp is a right angle, and Zp is the tangent at p, or the tangents at P and p intersect on the directrix.

PROP. X.

If ZP, Zp meet latus rectum produced in D and d, prove $SD = Sd$.

Proposition XI.

If the normal at P *intersects the transverse axis in* G,

SG = e . SP.

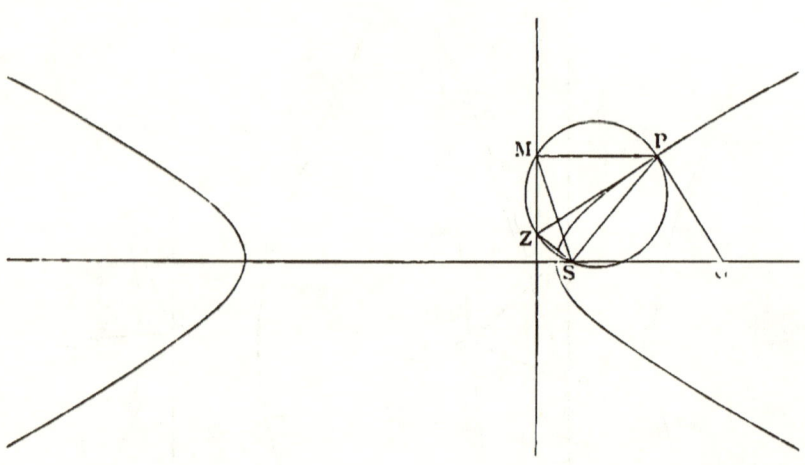

Draw the tangent *PZ*, join *SZ*, draw *PM* perpendicular to the directrix, and join *SM*.

ZMP and *ZSP* are right angles;   [Prop. 10.
therefore the circle, on *ZP* as diameter, passes through *M* and *S*.   [Euc. III. 31.

Since *ZPG* is a right angle, *PG* touches the circle.
        [Euc. III. 16.

Therefore the angle *SPG* = angle *SMP* in the alternate segment.   [Euc. III. 32.

Also angle *GSP* = angle *SPM*.   [Euc. I. 29.

Therefore the triangles *SPG*, *SMP* are similar;

∴   *SG* : *SP* = *SP* : *PM* ;

∴   *SG* = e . *SP*.

HYPERBOLA.

Proposition XII.

The tangent and normal to a hyperbola at any point P are respectively the internal and external bisectors of the angle between the focal distances.

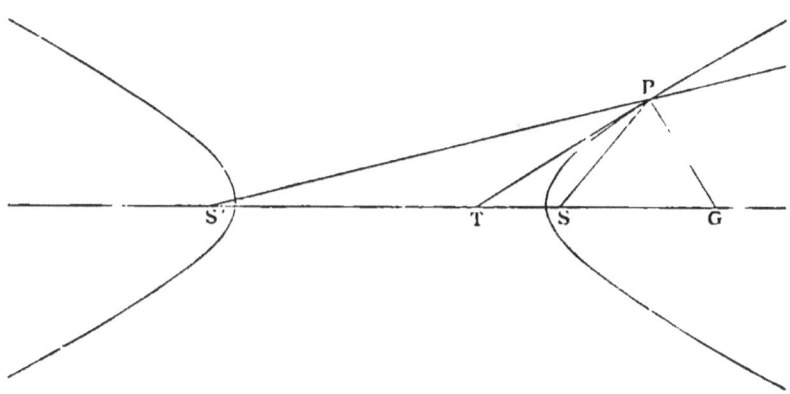

Let TP be the tangent and PG the normal, meeting the transverse axis in T and G.

$$SG = e \cdot SP, \qquad \text{[Prop. 11.}$$
and
$$S'G = e \cdot S'P;$$
$$\therefore SG : S'G = SP : S'P;$$

therefore PG bisects the angle SPS' externally. [Euc. VI. A.

Therefore the complements SPT, $S'PT$ are equal, and PT bisects the angle SPS' internally.

NOTE. Compare this with prop. 13 of the ellipse.

Prop. XII.

1. Given one focus of an hyperbola, one point and the tangent at the point, find the locus of the other focus.

2. If an ellipse and hyperbola have the same foci, they intersect at right angles.

HYPERBOLA.

PROPOSITION XIII.

The feet of the perpendiculars (SY, S'Y') from the foci on the tangent at P are on the circle described on AA' as diameter.

Also if CE, parallel to the tangent at P, intersects S'P in E, PE = CA.

Also $\qquad SY \cdot S'Y' = CB^2.$

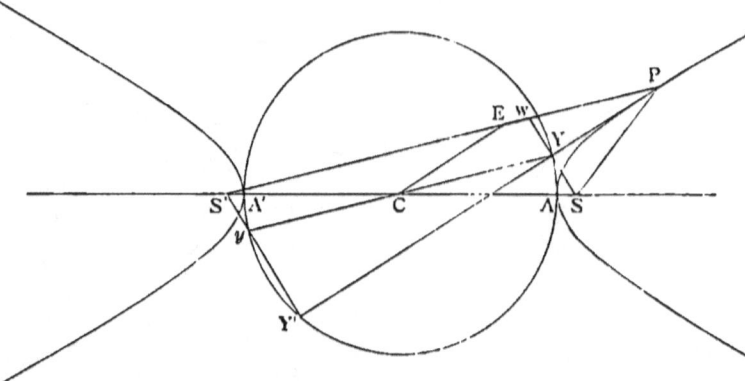

Produce SY to meet $S'P$ in W. Join CY.

In the triangles YPS, YPW, YP is common, right angles PYS, PYW are equal, angle $YPS =$ angle YPW; [Prop. 12.

$\therefore SY = YW, SP = PW$; [Euc. I. 26.

therefore $S'W$ is parallel to CY; [Euc. VI. 2.

$$\begin{aligned}\therefore CY &= \tfrac{1}{2}(S'W) \\ &= \tfrac{1}{2}(S'P - PS) \\ &= \tfrac{1}{2}AA' \qquad \text{[Prop. 7.}\\ &= CA\,;\end{aligned}$$

therefore Y is on the circle on AA' as diameter.

Similarly, Y' is on the auxiliary circle.

Also $YCEP$ is a parallelogram; therefore
$$PE = CY = CA.$$

Let $Y'S'$ meet the circle in y and join Yy.

Then, $YY'y$ being a right angle, Yy passes through the centre C, [Euc. III. 31.

$$\begin{aligned}SY &= S'y, \qquad \text{[Euc. I. 4.}\\ SY \cdot S'Y' &= S'y \cdot S'Y' \\ &= AS' \cdot S'A' \qquad \text{[Euc. III. 35.}\\ &= CB^2 \qquad \text{[Prop. 8.}\end{aligned}$$

Proposition XIV.

If the tangent at P meets the transverse axis in T,
$$CN \cdot CT = CA^2.$$

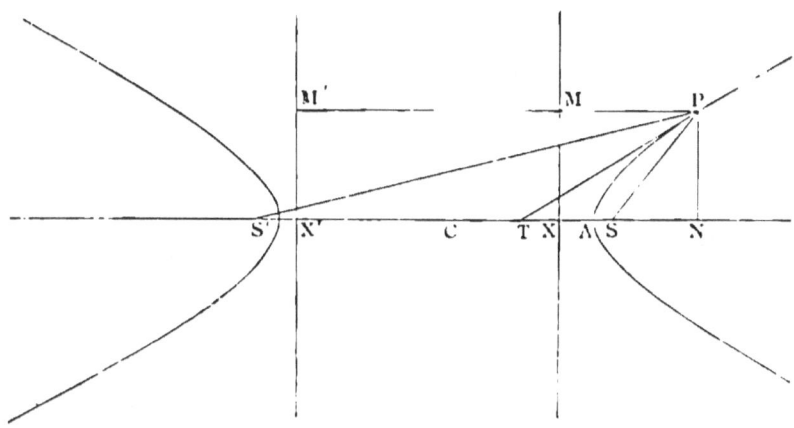

Draw PMM' perpendicular to the directrices.
Join SP, $S'P$.

Then, $\because PT$ bisects the angle SPS';   [Prop. 12.

$\therefore ST : S'T = SP : S'P$ [Euc. VI. A.
$ = PM : P'M$
$ = NX : NX'$;

$\therefore ST + S'T : S'T \sim ST = NX + NX' : NX' \sim NX$;

$\therefore 2CS : 2CT = 2CN : 2CX$;

$\therefore CN \cdot CT = CS \cdot CX$
$ = CA^2.$ [Prop. 6.

Prop. XIII.
The riders on page 52 are also true for the hyperbola.

Prop. XIV.
1. Prove prop. 16 of the ellipse by this method.
2. If Tp be drawn perpendicular to the axis to meet the auxiliary circle in p, prove that Np is a tangent to the circle.
3. Prove $CN \cdot NT = AN \cdot NA'$.

HYPERBOLA.

Proposition XV.

If the tangent at P meets the conjugate axis produced in t, and Pn is the perpendicular from P on the conjugate axis,

$$Cn \cdot Ct = CB^2.$$

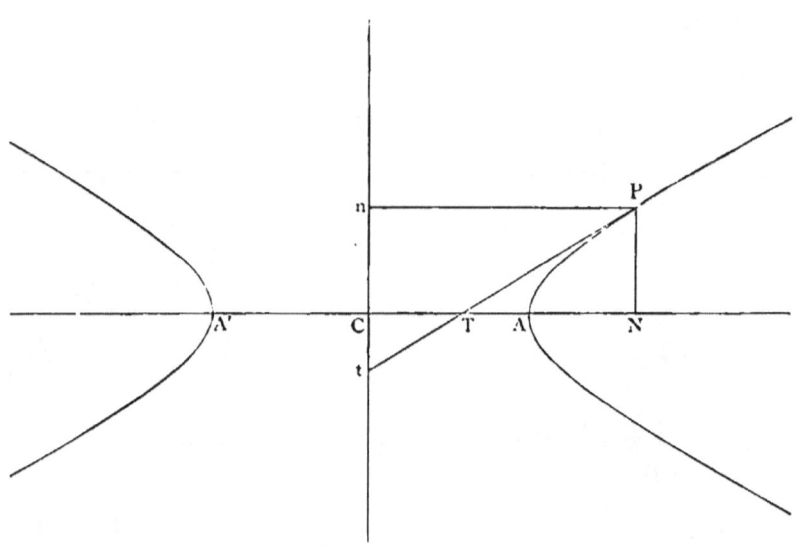

Draw the ordinate PN.
Then, by similar triangles,
$$TN : CT = PN : Ct.$$
$$\therefore TN \cdot CN : CN \cdot CT = PN^2 : Ct \cdot PN;$$
$$\therefore TN \cdot CN : CA^2 = PN^2 : Ct \cdot Cn. \quad \text{[Prop. 14.}$$
But $\quad TN \cdot CN = CN^2 - CT \cdot CN$
$$= CN^2 - CA^2 \quad \text{[Prop. 14.}$$
$$= AN \cdot A'N; \quad \text{[Euc. II. 5.}$$
$$\therefore AN \cdot A'N : CA^2 = PN^2 : Ct \cdot Cn.$$
Therefore, alternately,
$$AN \cdot A'N : PN^2 = CA^2 : Ct \cdot Cn.$$
But $\quad AN \cdot A'N : PN^2 = CA^2 : CB^2, \quad \text{[Prop. 3.}$
$$\therefore Ct \cdot Cn = CB^2.$$

HYPERBOLA.

PROPOSITION XVI.

If PF *is the perpendicular from* P *on a line through* C *parallel to the tangent at* P, *and if the normal at* P *meets the conjugate axis in* g, *then*

$$PF \cdot PG = CB^2 \text{ and } PF \cdot Pg = CA^2.$$

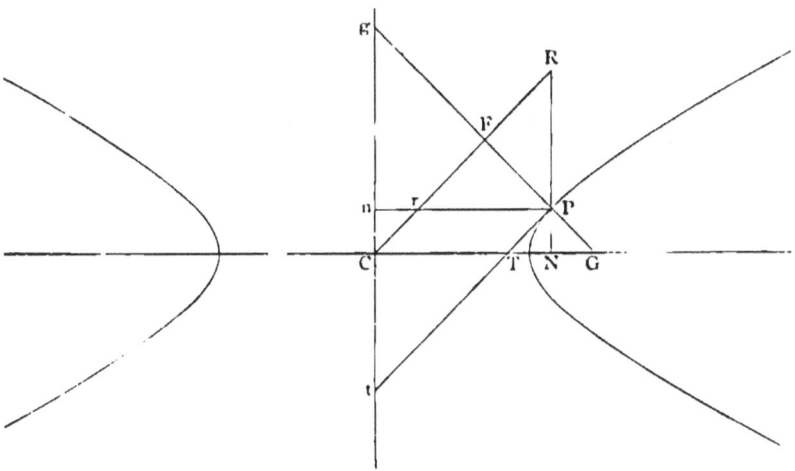

Draw RPN, Prn, perpendiculars on the axes meeting CF in R and r, and let the tangent at P meet the axes in T and t.

Then since the angles at N and F are right angles, therefore a circle passes round $GNFR$.   [Euc. III. 22.

Therefore $PG \cdot PF = PN \cdot PR$ [Euc. III. 35.

$$= Cn \cdot Ct = CB^2. \quad \text{[Prop. 15.}$$

Again, because the angles at F and n are right angles, therefore a circle passes round $gFrn$:

$$\therefore PF \cdot Pg = Pn \cdot Pr \quad \text{[Euc. III. 36.}$$
$$= CN \cdot CT = CA^2. \quad \text{[Prop. 14.}$$

NOTE. It will be seen afterwards that the line CFR, referred to in the enunciation, is the diameter CD conjugate to CP.

Proposition XVII.

$NG : CN = CB^2 : CA^2$ and $CG = e^2 . CN$.

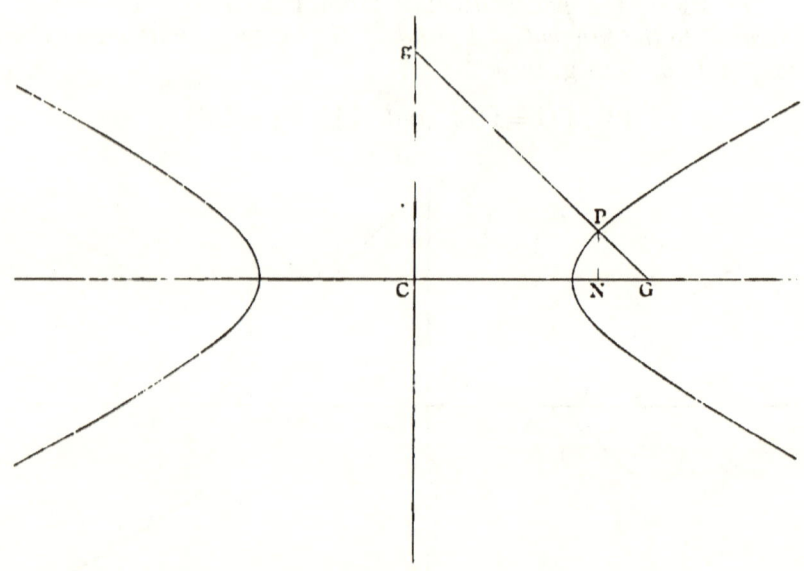

Produce GP to meet the conjugate axis in g.

Then $\quad NG : CN = PG : Pg \quad$ [Euc. VI. 2.
$\quad\quad\quad\quad = PG . PF : Pg . PF$
$\quad\quad\quad\quad = CB^2 : CA^2.$ [Prop. 16.

Again, since $\quad NG : CN = CB^2 : CA^2;$

$\therefore\ CN + NG : CN = CA^2 + CB^2 : CA^2;$

$\therefore\ CG : CN = CS^2 : CA^2 \quad$ [Prop. 8.
$\quad\quad\quad = e^2 : 1.$ [Prop. 6.

$\therefore\ CG = e^2 . CN.$

Prop. XVII.

1. Prove that $\quad CG . Cn : Cg . CN = BC^2 : AC^2.$

2. In the R. H. prove $\quad$ (a) $CN = NG$,
$\quad\quad\quad\quad\quad\quad\quad\quad$ (b) $PG = Pg = CP.$

HYPERBOLA.

Proposition XVIII.

If from any point O *on the tangent at* P, OI *is drawn perpendicular to the directrix, and* OU *perpendicular to* SP, *then* SU = e . OI (Adams's property).

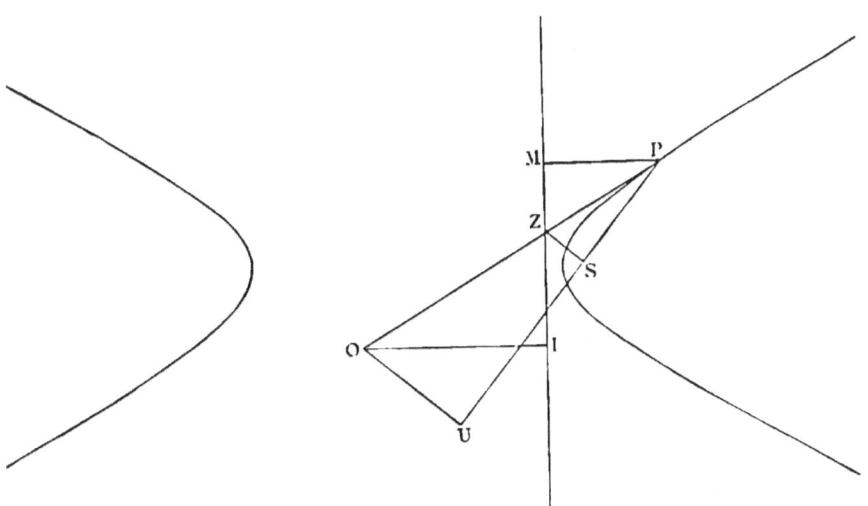

Join SZ, and draw PM perpendicular to the directrix.

Then since the angle ZSP is a right angle, ZS is parallel to OU.

$$\therefore SU : SP = ZO : ZP$$
$$= OI : MP.$$
$$\therefore SU : OI = SP : MP$$
$$= e : 1.$$
$$\therefore SU = e . OI.$$

If O be a point on the tangent, such that OQQ', drawn perpendicular to the transverse axis, meets the curve in Q and Q', then $SU = SQ$ and $OU^2 = OQ . OQ'$. See ellipse prop. 20, figure 2.

Proposition XIX.

To draw a pair of tangents OQ, OQ' to a hyperbola from an external point O.

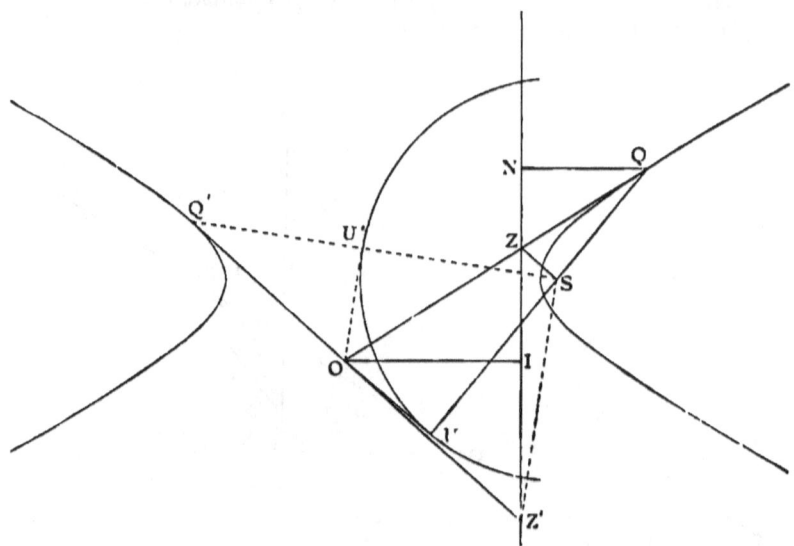

Draw OI perpendicular to the directrix. With centre S and radius $e.OI$ describe a circle, and draw OU, OU' tangents to it from O.

Draw SZ perpendicular to SU meeting the directrix in Z. Join ZO and produce it to meet SU in Q. Draw QN perpendicular to the directrix.

Then
$$SQ : SU = QZ : OZ$$
$$= QN : OI;$$
$$\therefore SQ : QN = SU : OI = e : 1;$$
therefore Q is on the hyperbola.

And since QSZ is a right angle, therefore OQ is the tangent to the hyperbola at Q. [Prop. 10.

So by drawing SZ' perpendicular to SU', and joining OZ' and producing it to meet SU' in Q', OQ' is the other tangent.

Note. This problem is solved by the principles of Proposition 18, but a construction could also be founded on Propositions 12 or 13.

HYPERBOLA.

Proposition XX.

Tangents OQ, OQ' subtend equal or supplementary angles OSQ, OSQ' at the focus S according as Q, Q' are on the same or opposite branches of the hyperbola.

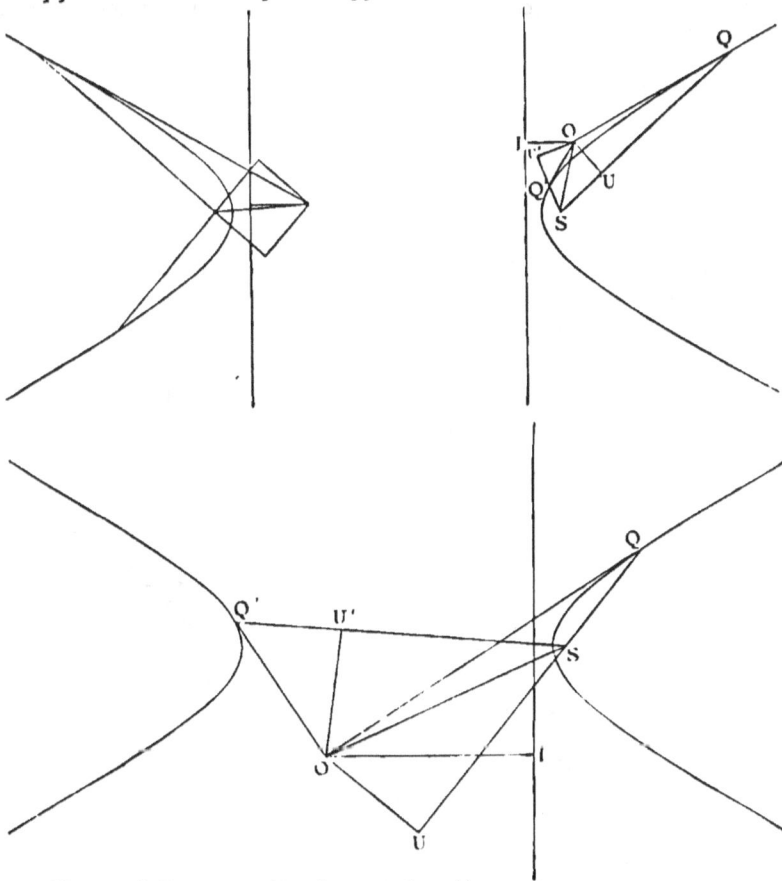

Draw OI perpendicular to the directrix.

Join OS, SQ, SQ', and draw OU, OU' perpendiculars on SQ, SQ'.

Then $SU = e \cdot OI = SU'$. [Prop. 18.

Therefore the triangles OSU, OSU' are equal in all respects. [Euc. I. 26.

Therefore the angle $OSU =$ angle OSU'.

Therefore, in fig. 1, angle $OSQ =$ angle OSQ';

And, in fig. 2, angles OSQ, OSQ' are supplementary angles

Note. If O lies between the directrices, use the left-hand part of fig. 1.

100   HYPERBOLA.

Prop. XX.

1. The portion of any tangent intercepted between the tangents at the vertices subtends a right angle at either focus.

2. The locus of the centre of the inscribed circle of the triangle SPS' is a straight line.

3. In any conic the chord of contact QQ' is divided harmonically by SO and the directrix.

Proposition XXI.

OQ, OQ' *are inclined at equal or supplementary angles to* OS, OS' *according as* Q, Q' *are on opposite or the same branches of the hyperbola.*

Case 1. Join SQ, SQ', $S'Q$, $S'Q'$, and produce QS to W, and let SQ' meet $S'Q$ in K.

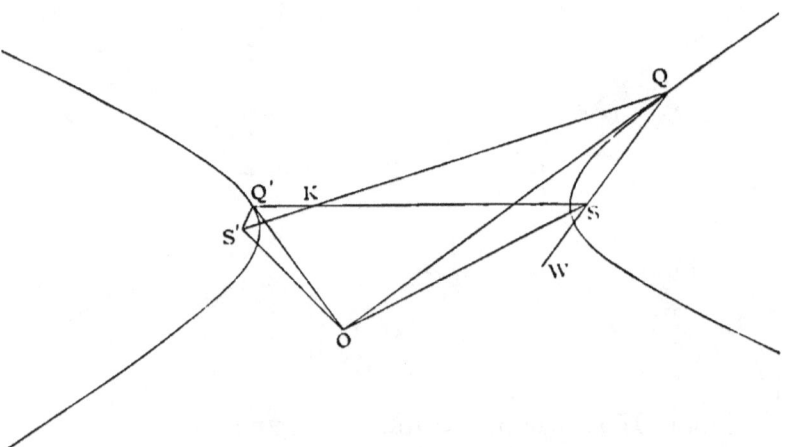

Then,   angle $SOQ = OSW - OQS$ [Euc. I. 32.

$\qquad\qquad = \tfrac{1}{2}Q'SW - \tfrac{1}{2}S'QS$ [Props. 20, 12.

$\qquad\qquad = \tfrac{1}{2}SKQ.$ [Euc. I. 32.

Similarly, $S'OQ' = \tfrac{1}{2}S'KQ'$;

$\therefore\ SOQ = S'OQ'.$

HYPERBOLA.   101

Case 2.

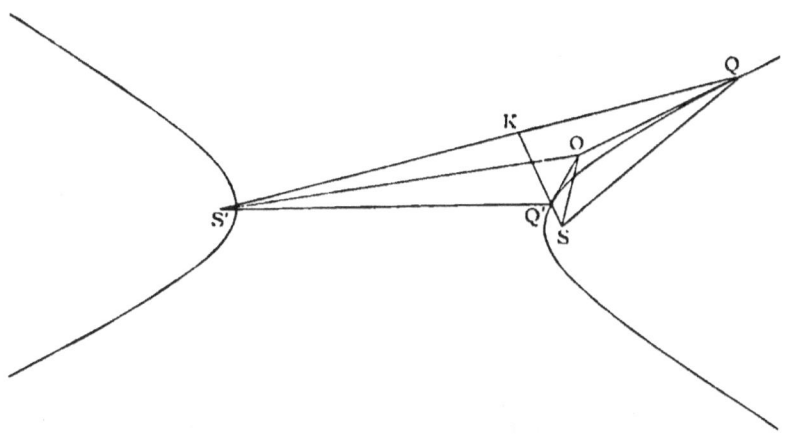

$$SOQ = 180° - OSQ - OQS \qquad \text{[Euc. I. 32.}$$
$$= 180° - \tfrac{1}{2}QSQ' - \tfrac{1}{2}SQS'' \qquad \text{[Props. 20, 12.}$$
$$= 180° - \tfrac{1}{2}SKS''. \qquad \text{[Euc. I. 32.}$$

Again, $\quad S'OQ' = 180° - OQ'S' - OS'Q' \qquad \text{[Euc. I. 32.}$
$$= \tfrac{1}{2}SQ'S' - \tfrac{1}{2}QS''Q' \qquad \text{[Props. 12, 20.}$$
$$= \tfrac{1}{2}SKS''; \qquad \text{[Euc. I. 32.}$$
$$\therefore SOQ = 180° - S'OQ'.$$

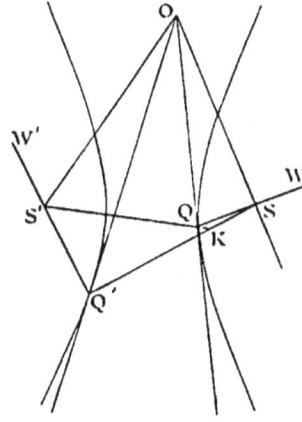

In Case 2 the point O lies within one of the two angles between the asymptotes, which contain the two branches of the hyperbola; in Case 1 O lies within one of the other two angles between the asymptotes.

Also the nature of the proof depends slightly upon whether O lies between the directrices or not. For Case 1 in the text the point O is between the directrices; in this figure it is not so, and K consequently lies in $S'Q$ produced.

Again, the two positions of O, given in prop. 20, figure 1, will supply opposite examples of Case 2.

HYPERBOLA.

Def. A hyperbola which has CB and CA for transverse and conjugate axes respectively is called the *conjugate hyperbola*.

Note. The conjugate hyperbola has the same asymptotes as the original hyperbola, because they are diagonals of the same rectangle. [Prop. 4.

Proposition XXII.

If through any point P *on the curve a line be drawn parallel to* CA *or* CB, *meeting the asymptotes in* p, p', *the rectangle* Pp . Pp' *is* = *to the square on* CA *or* CB *respectively. The same is true if* P *be on the conjugate hyperbola.*

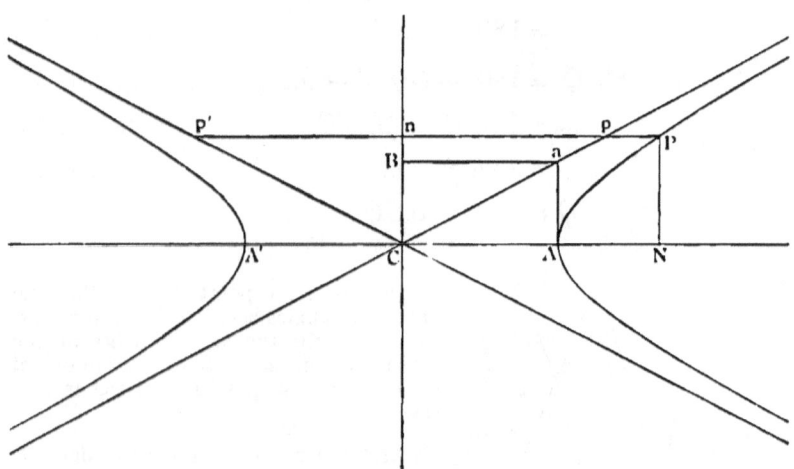

Case 1. Draw Ppp' parallel to CA, meeting CB in n.

Then $\quad PN^2 : CN^2 - CA^2 = CB^2 : CA^2;\quad$ [Prop. 3.

$\therefore\ Cn^2 : Pn^2 - CA^2 = CB^2 : CA^2.$

Also $Cn^2 : pn^2 = CB^2 : Ba^2 = CB^2 : CA^2$;

$\therefore Pn^2 - CA^2 = pn^2$;

$\therefore Pn^2 - pn^2 = CA^2$;

or $Pp \cdot Pp' = CA^2$.

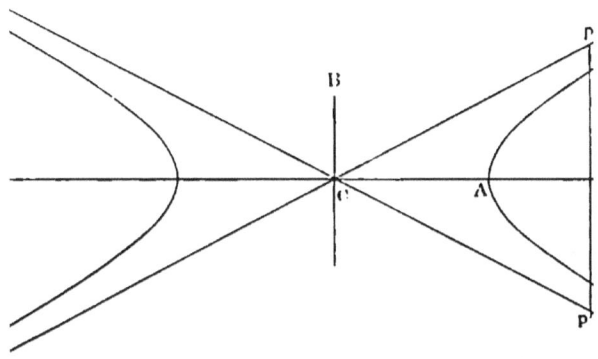

Case 2. Draw Ppp' parallel to CB.

Then $\qquad Pp \cdot Pp' = CB^2$.

Cases 3 *and* 4.

Since it has been proved for both axes of the hyperbola that
$$Pp \cdot Pp' = CA^2 \text{ or } CB^2 \text{ respectively,}$$
therefore it is also true if P be on the conjugate hyperbola, as in the figures below.

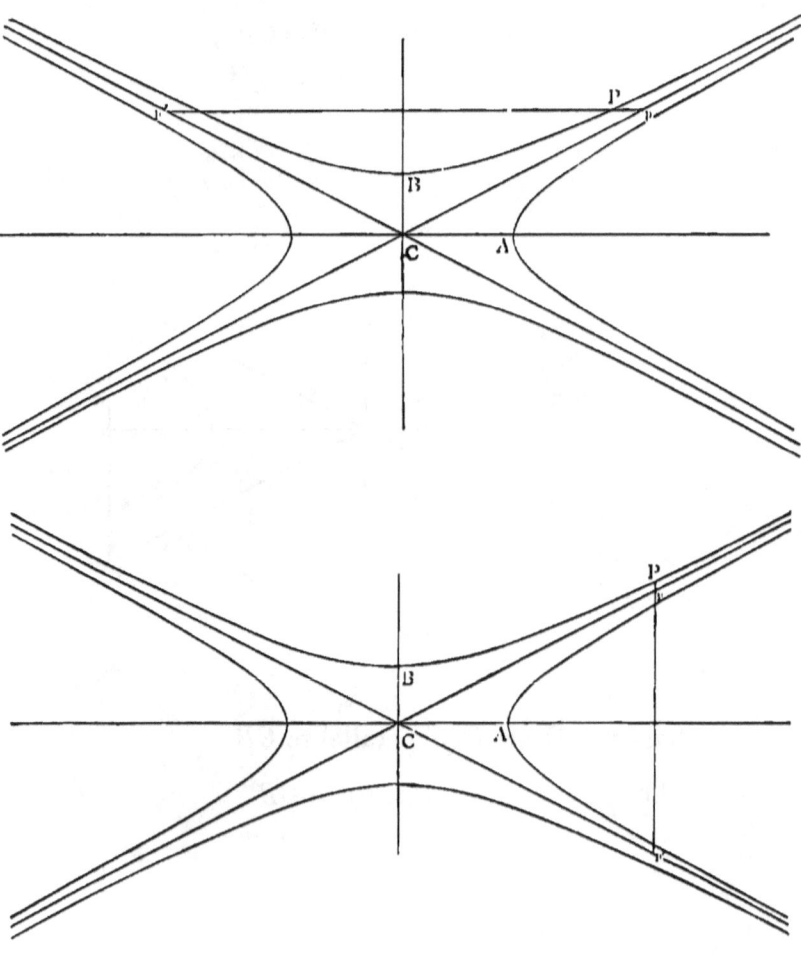

Prop. XXIII.

QQ' is a chord of a hyperbola parallel to the tangent at P. Pp, Qq, $Q'q'$ are drawn parallel to one asymptote and terminated by the other.

Prove $Cq \cdot Cq' = Cp^2$.

Proposition XXIII.

If through any two points P, Q *on the curve or its conjugate two parallel straight lines be drawn to meet the asymptotes in* p, p'; q, q' *respectively, the rectangle*
$$Pp \cdot Pp' = Qq \cdot Qq'.$$

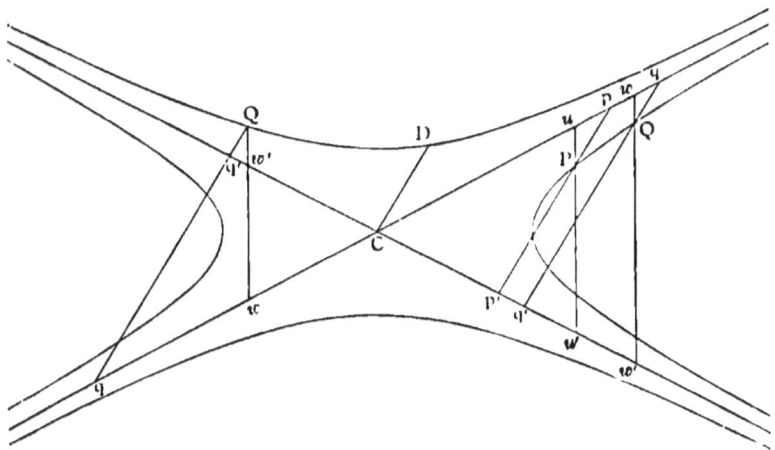

First let P and Q be on the same branch of the hyperbola.

Through P and Q draw lines parallel to CB meeting the asymptotes in u, u'; w, w'.

By similar triangles,
$$Pp : Pu = Qq : Qw,$$
and
$$Pp' : Pu' = Qq' : Qw'.$$

Therefore, by multiplying,
$$Pp \cdot Pp' : Pu \cdot Pu' = Qq \cdot Qq' : Qw \cdot Qw'.$$
But $\quad Pu \cdot Pu' = CB^2 = Qw \cdot Qw';$ [Prop. 22.
$$\therefore Pp \cdot Pp' = Qq \cdot Qq'.$$

The same argument applies whether Q be on the hyperbola or its conjugate; both cases are shown on the figure.

Note. Through the centre draw CD parallel to Qq or Pp, meeting the curve or its conjugate at D, then applying this proposition to the points Q and D,
$$Qq \cdot Qq' = DC \cdot DC = CD^2.$$

Proposition XXIV.

If any straight line cut the curve in Q, Q', *and the asymptotes in* qq', Qq = Q'q';

And if the tangent rPr' *meet the asymptotes in* r *and* r', *then* Pr = Pr'.

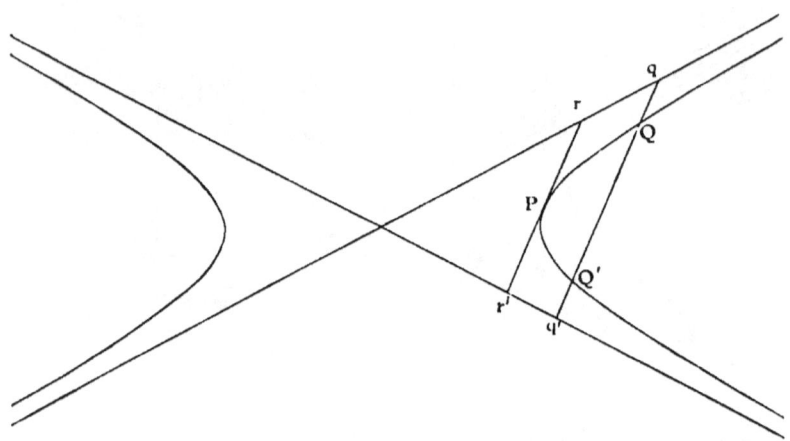

$$Qq \cdot Qq' = Q'q' \cdot Q'q; \qquad \text{[Prop. 23.}$$
$$\therefore Qq \cdot QQ' + Qq \cdot Q'q' = Q'q' \cdot QQ' + Q'q' \cdot Qq;$$
$$\therefore Qq \cdot QQ' = Q'q' \cdot QQ';$$
$$\therefore Qq = Q'q'.$$

Let QQ' move parallel to itself until it becomes the tangent at P.

Since $Qq = Q'q'$ always;
$$\therefore Pr = Pr'.$$

NOTE. QQ' may be on opposite branches of the hyperbola, in this case there is not a tangent to this hyperbola parallel to QQ'.

Prop. XXIV.

1. The same is true if qq' be on the conjugate hyperbola.

2. If the normal at P meet the axes in G, g; G, g, r, r' lie on a circle passing through the centre.

HYPERBOLA.

Proposition XXV.

The locus of the middle points of a system of parallel chords is a straight line passing through the centre;

And the tangent at either end of the straight line is parallel to the chords.

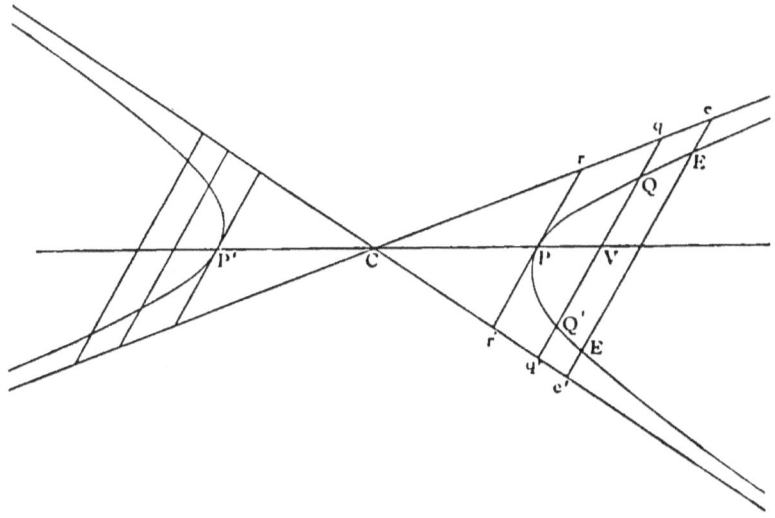

Let QQ', EE', &c. be a system of parallel chords meeting the asymptotes in q, q'; e, e'; &c.

Draw CV bisecting QQ' in V.

Then CV also bisects qq', because $Qq = Q'q'$. [Prop. 24.

Therefore, by similar triangles, CV bisects ee'.

Therefore it bisects EE'; because $Ee = E'e'$. [Prop. 24.

Therefore CV bisects all chords parallel to QQ'.

Let CV meet the curve in P, and let QQ' move parallel to itself towards P.

Then, since QQ' is always bisected by CPV, Q and Q' ultimately coincide with P; therefore the tangent at P is parallel to the system of parallel chords bisected by CPV.

HYPERBOLA.

DEF. A straight line (*CP*) passing through the middle points of a system of parallel chords is called a *diameter*.

DEF. A straight line (*QV*) drawn from any point on the curve parallel to the tangent at the extremity of the diameter (*PCP'*) is called *the ordinate to the diameter*.

N.B. If the diameter is the transverse axis, the ordinate has the usual meaning.

NOTE. The *length* of that portion of a diameter, which is intercepted by the hyperbola or its conjugate, is sometimes called the *diameter*.

PROPOSITION XXVI.

If one diameter bisects chords parallel to a second, then the second diameter bisects chords parallel to the first.

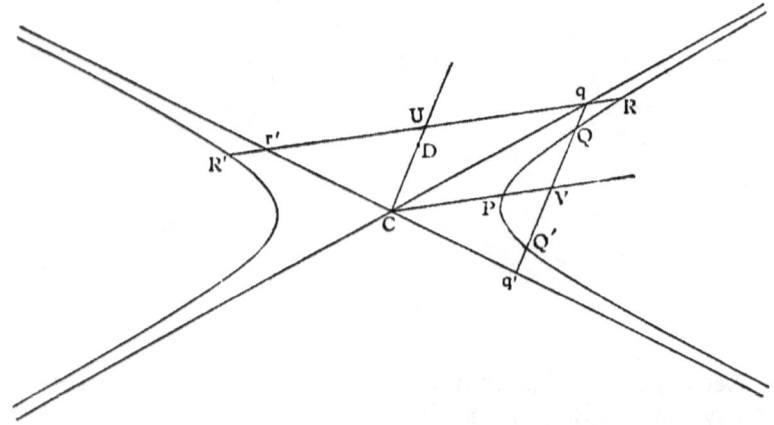

Let *CP* bisect *QQ'* in *V* and draw *CD* parallel to *QQ'*.

Produce *QQ'* to meet the asymptotes in *q, q'*.

Through *q* draw *RqUr'R'* parallel to *CP*, meeting the curve in *R* and *R'*, and the asymptotes in *q, r'*, and *CD* in *U*.

Then, because *Qq* = *Q'q'*, therefore *qq'* is bisected in *V*; and *CV* is parallel to *qr'*,

∴ *Cr'* = *Cq'* ; [Euc. VI. 2.
∴ *r'U* = *Uq* ; [Euc. VI. 2.

and *Rq* is equal to *R'r'*,

∴ *R'U* = *RU* ; [Prop. 24.

therefore *CD* bisects all chords parallel to *CP*. [Prop. 25.

Proposition XXVI. (*Aliter.*)

If one diameter bisects chords parallel to a second, then the second diameter bisects chords parallel to the first.

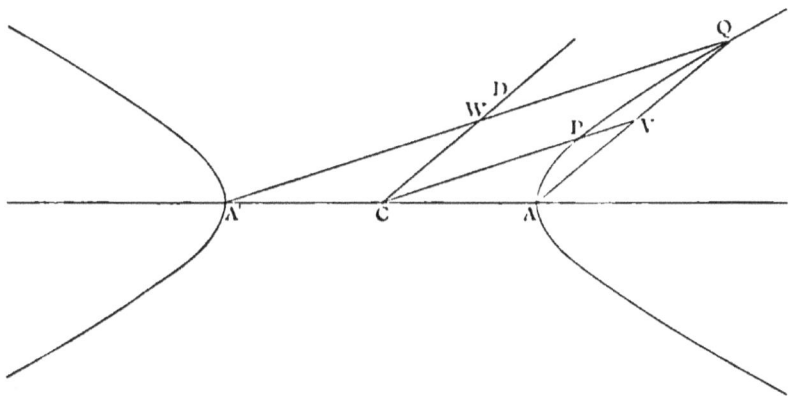

Draw AQ parallel to CD, meeting CP in V.

Join $A'Q$ cutting CD in W.

Since AQ is bisected in V and AA' in C; therefore $A'Q$ is parallel to CP.

And because CD is parallel to AQ, therefore $A'Q$ is bisected in W.

Therefore CD bisects the chord $A'Q$ parallel to CP.

Therefore CD bisects all chords parallel to CP.

Def. If two diameters are so related that each bisects chords parallel to the other, they are called *conjugate diameters*.

Note. Of two conjugate diameters one will meet the hyperbola, and the other the conjugate hyperbola.

HYPERBOLA.

Def. Chords (QP, QP') which join any point (Q) on a hyperbola to the extremities of a diameter (PCP') are called *supplemental chords*.

Proposition XXVII.

Supplemental chords are parallel to conjugate diameters.

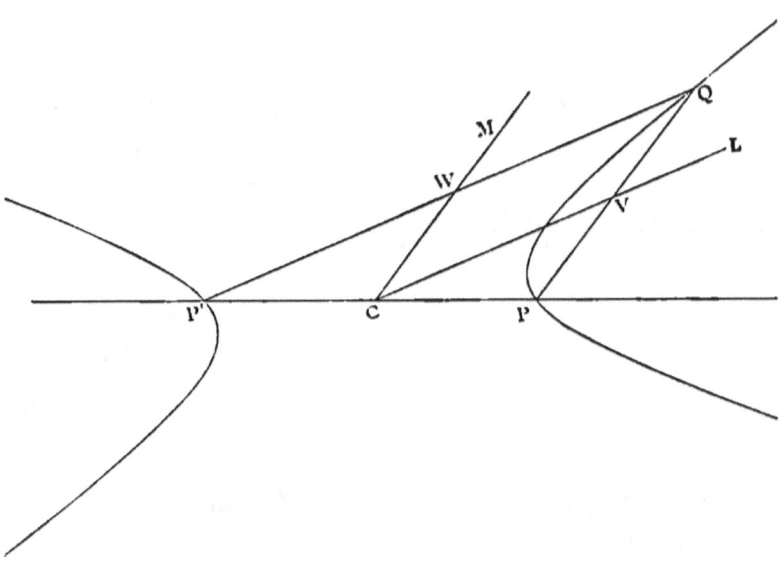

Draw the diameters CL, CM parallel to the supplemental chords $P'Q$, PQ cutting them in W and V.

Then $\qquad PV : VQ = PC : CP'$; [Euc. VI. 2.

$\qquad \therefore PV = VQ$;

$\therefore$ CL bisects PQ, and all other chords parallel to CM.
[Prop. 25.

Similarly CM bisects all chords parallel to CL; therefore CL, CM are conjugate diameters.

HYPERBOLA.

PROPOSITION XXVIII.

Tangents to the hyperbola and its conjugate at their intersections with conjugate diameters PCP', DCD' form a parallelogram whose angular points are on the asymptotes.

Also PD is bisected by one asymptote and is parallel to the other.

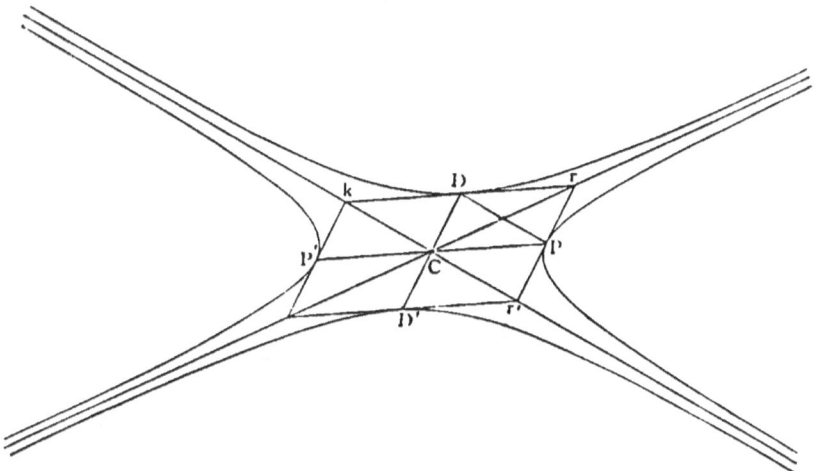

Draw the tangent rPr' meeting the asymptotes in r and r'.
Join CD.
Then since CD is conjugate to CP,
∴ CD is parallel to rr'.
Therefore, by Prop. 23, observing that DC meets both the asymptotes in C,
$$DC^2 = Pr . Pr' = Pr^2;\qquad\text{[Prop. 24.}$$
∴ $DC = Pr$ and is parallel to it;
∴ rD is parallel to CP; [Euc. I. 33.
∴ rD is the tangent at D. [Prop. 25.

Similarly the tangents at D and P' meet on the asymptotes, and the four tangents form a parallelogram with its angular points on the asymptotes.

Join PD, and let rD meet the other asymptote in k.
Then $\qquad rP = Pr'$,
and $\qquad rD = Dk;$
∴ PD is parallel to kr',
and $CPrD$ is a parallelogram,
∴ PD is bisected by the asymptote.

For riders see page 113.

Proposition XXIX.

Straight lines through P *and* D *parallel to the axes form a rectangle with two angular points on one of the asymptotes.*

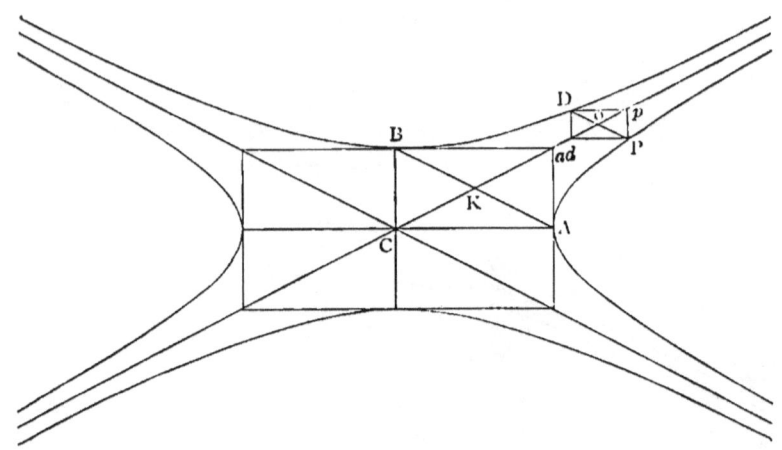

Draw Pp parallel to CB, meeting the asymptote in p; and join pD.

Let AB, PD intersect the asymptote at k and o, then AB and PD are both bisected by the asymptote, and they are parallel to one another (Prop. 28);

Hence poP, aKA are similar triangles.

∴ $Pp : Aa = Po : Ak$
   $= PD : AB.$ [Prop. 28.

And angle pPD = angle aAB.

Therefore the triangles pPD, aAB are similar.
[Euc. vi. 6.

Therefore pD is parallel to aB, i.e. to CA.

Similarly, if Dd be drawn parallel to CB,
Then Pd is parallel to CA.

HYPERBOLA.

Proposition XXX.

$$CP^2 \sim CD^2 = CA^2 \sim CB^2.$$

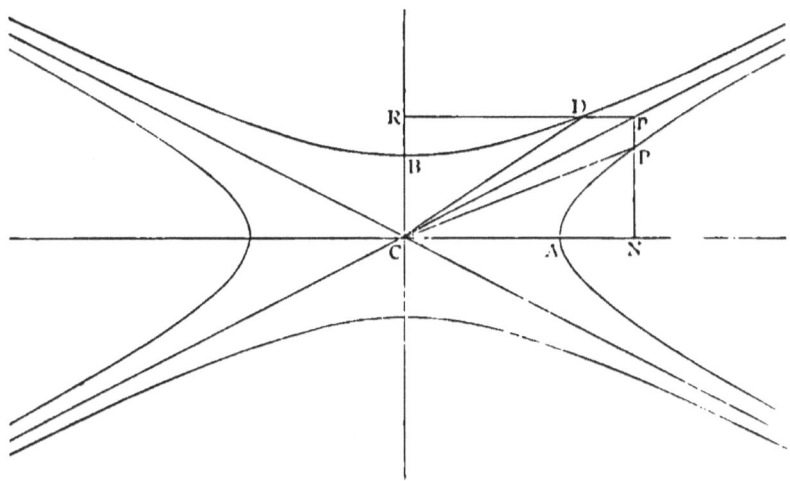

Draw the ordinates PN, DR to the axes and produce them to meet in p, then p lies on the asymptote (Prop. 29).

Then
$$CB^2 = pN^2 - PN^2 \qquad \text{[Prop. 24.}$$
$$= Cp^2 - CP^2. \qquad \text{[Euc. I. 47.}$$
Also
$$CA^2 = pR^2 - DR^2 \qquad \text{[Prop. 24.}$$
$$= Cp^2 - CD^2; \qquad \text{[Euc. I. 47.}$$
$$\therefore CA^2 \sim CB^2 = CP^2 \sim CD^2.$$

Prop. XXVIII.

In the R. H. prove

1. $CP = CD$ and the asymptotes bisect the angle between any pair of conjugate diameters.
2. CP and CD make complementary angles with the axes.
3. Diameters at right angles are equal.
4. The angle between any two diameters is equal to the angle between their conjugates.
5. The angles subtended by any chord at the extremities of a diameter PP' are equal or supplementary.
6. If a R. H. circumscribe a triangle, the locus of the centre is the nine-point circle.

C. G.   8

Proposition XXXI.

If any tangent rPr′ *to the hyperbola meet the asymptotes in* r *and* r′, *the parallelogram* CPrD *is constant,*

(*or* PF . CD = AC . BC).

Also the triangle rCr′ *is constant.*

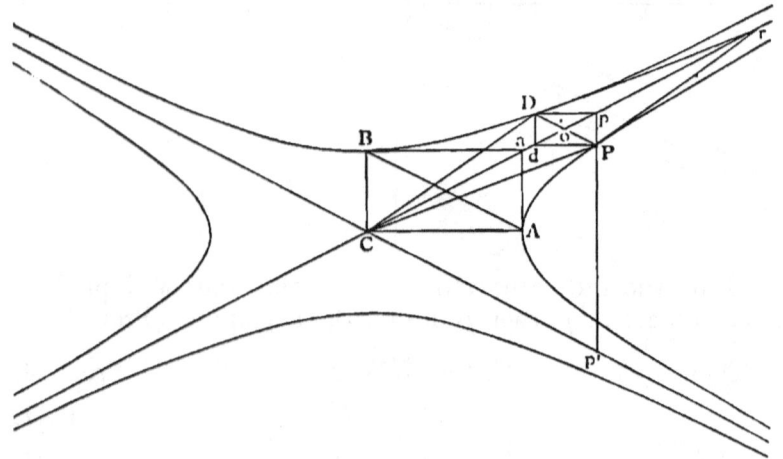

Draw Aa, Ba parallel to the axes, meeting the asymptote in a.

Draw the double ordinate through P meeting the asymptotes in p, p′.

Complete the parallelogram DpPd. Join DP cutting the asymptote in o. Join AB.

Then △ DCP : △ DpP = Co : op
     = p′P : Pp.   [Euc. VI. 2.

Again, △ BCA : △ DpP = BC² : Pp²   [Euc. VI. 19.
     = Pp . Pp′ : Pp²   [Prop. 22.
     = Pp′ : Pp ;

∴ triangle DCP = triangle BCA.

∴ parallelogram $CPrD$ = parallelogram $CAaB$, which is constant.

Or $\qquad PF . CD = AC . BC.$ [See fig. of Prop. 16.

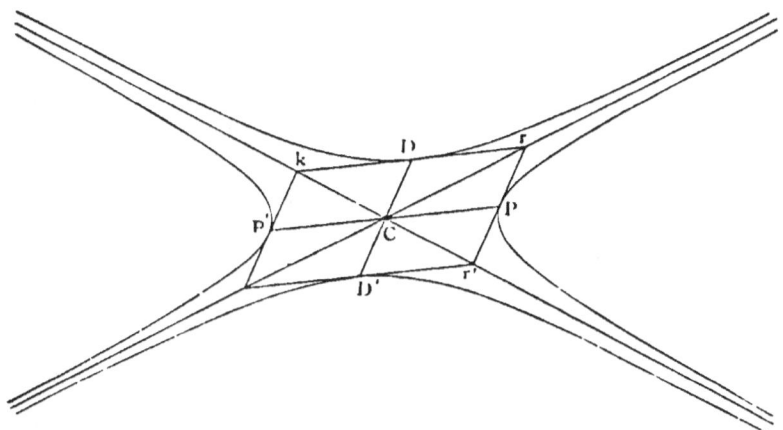

Also the triangle rCr' = parallelogram $CPrD$, for they are, each of them, a quarter of the parallelogram formed by the tangents at P, D, P', D'.

Therefore the triangle rCr' is constant.

Prop. XXXI.

1. If Po, Po' be drawn respectively parallel to one asymptote and terminated by the other, $Po . Po' = \frac{1}{4}CS^2$.

2. If the two asymptotes and a point on the curve be given in position, find the axes and foci.

3. Two tangents to an hyperbola meet the asymptotes in R, r, T, t respectively. Prove Rt parallel to rT.

4. In the R. H. if CZ be drawn perpendicular to the tangent at P, prove that $CZ . CP = CA^2$.

Proposition XXXII.

QV *is an ordinate of the diameter* PCP', CD *the diameter parallel to* QV. *Then*
$$QV^2 : PV \cdot P'V = CD^2 : CP^2.$$

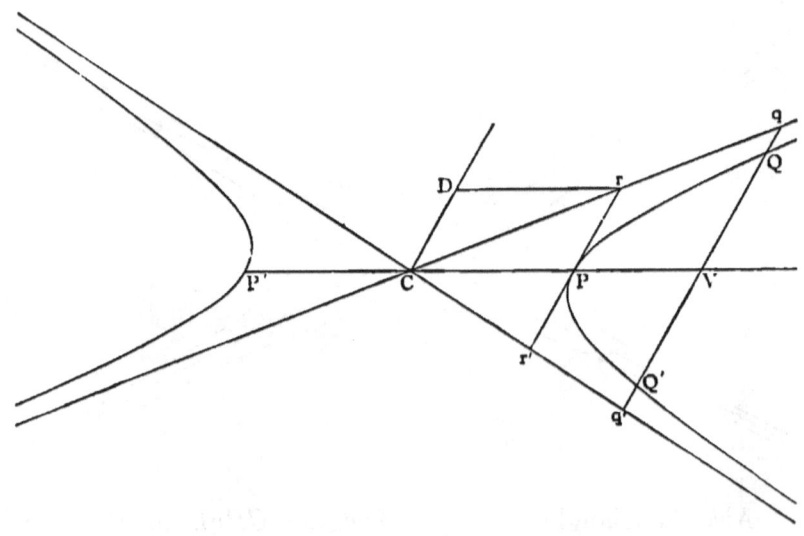

Let QV meet the asymptotes in q, q'. Draw the tangents at P, D, meeting the asymptotes in r.   (Prop. 28.)

Then $\qquad CD^2 = Qq \cdot Qq' \qquad$ [Prop. 23.
$$= qV^2 - QV^2;$$
$$\therefore QV^2 = qV^2 - CD^2.$$

Also $\qquad PV \cdot P'V = CV^2 - CP^2.$

But, by similar triangles, CPr, CVq;
$$CV^2 - CP^2 : CP^2 = qV^2 - Pr^2 : Pr^2$$
$$= qV^2 - CD^2 : CD^2;$$
$$\therefore PV \cdot P'V : CP^2 = QV^2 : CD^2.$$

Alternando. $\quad QV^2 : PV \cdot P'V = CD^2 : CP^2.$

In the R. H. $QV^2 = PV \cdot P'V$.

Proposition XXXIII.

Tangents at the ends of any chord meet on the diameter which bisects the chord.

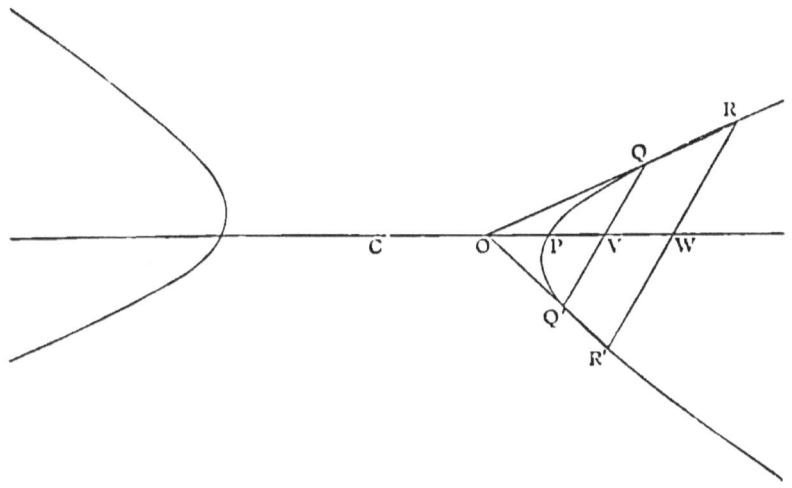

Let QQ', RR' be two parallel chords, join RQ, $R'Q'$ and produce them to meet in O.

Bisect QQ' in V, and let OV produced meet RR' in W.

By similar triangles,
$$QV : RW = OV : OW$$
$$= Q'V : R'W,$$
but $\qquad QV = Q'V,$
$$\therefore RW = R'W.$$

Since VW bisects the parallel chords QQ', RR' it is a diameter passing through the centre C. [Prop. 25.

Let R, R' move up to and ultimately coincide with Q, Q'; then OQR, $OQ'R'$ become a pair of tangents at Q, Q', and they still intersect on the diameter CV.

In any conic if a diameter meets the directrix in Z, SZ is perpendicular to the chords bisected by the diameter.

Proposition XXXIV.

QV *is an ordinate of the diameter* CP; *if the tangent at* Q *meets* CP *in* O, *then*
$$CV \cdot CO = CP^2.$$

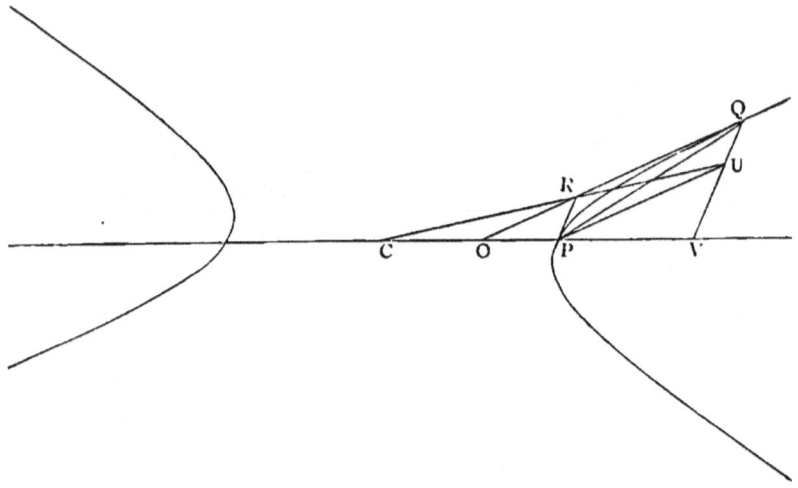

Draw PU parallel to OQ, and PR parallel to QV, and join PQ.

Then PR touches the hyperbola. [Prop. 25.

$RPUQ$ is a parallelogram; therefore RU bisects PQ; therefore RU passes through the centre C. [Prop. 33.

Now $\quad\quad CO : CP = CR : RU \quad\quad$ [Euc. vi. 2.
$\quad\quad\quad\quad\quad\quad = CP : CV,\quad\quad$ [Euc. vi. 2.
therefore $\quad\quad CP^2 = CO \cdot CV.$

Prop. XXXV.

1. If a r.r. circumscribe a triangle, it also passes through the orthocentre.

2. If OR be drawn parallel to an asymptote to meet the curve in R, and OPP' parallel to a fixed line to meet the curve in P, P', the rectangle $OP \cdot OP'$ varies as OR.

[See also riders on Prop. 34 of Ellipse.]

HYPERBOLA. 119

Proposition XXXV.

If two chords of a hyperbola intersect, the rectangles contained by their segments are as the squares of the parallel semi-diameters.

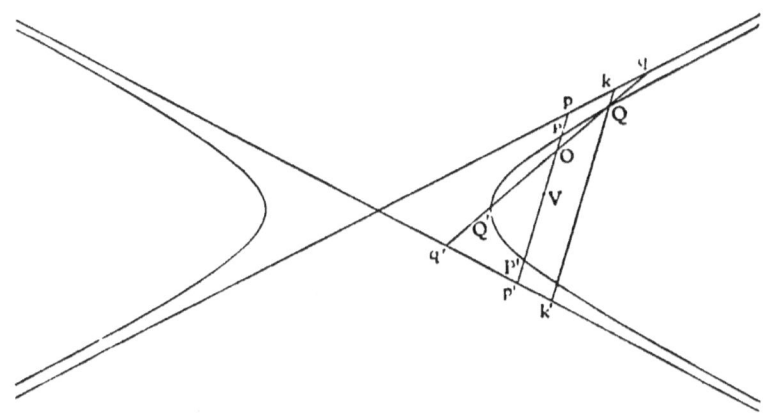

Let the chords POP', QOQ' meet the asymptotes at pp', qq'. Bisect PP' at V. Draw kQk' parallel to pp'.

Then
$$pO \cdot Op' = pV^2 - OV^2,\quad \text{[Euc. II. 5.}$$
$$PO \cdot OP' = PV^2 - OV^2;\quad \text{[Euc. II. 5.}$$
$$\therefore pO \cdot Op' - PO \cdot OP' = pV^2 - PV^2$$
$$= pP \cdot Pp';\quad \text{[Euc. II. 5.}$$
$$\therefore pO \cdot Op' - pP \cdot Pp' = PO \cdot OP'.$$

Similarly, $qO \cdot Oq' - qQ \cdot Qq' = QO \cdot OQ'$.

By similar triangles,
$$pO : qO = kQ : qQ,$$
and
$$Op' : Oq' = Qk' : Qq';$$
$$\therefore pO \cdot Op' : qO \cdot Oq' = kQ \cdot Qk' : qQ \cdot Qq'$$
$$= pP \cdot Pp' : qQ \cdot Qq';\quad \text{[Prop. 23.}$$
$$\therefore pO \cdot Op' - pP \cdot Pp' : qO \cdot Oq' - qQ \cdot Qq'$$
$$= pP \cdot Pp' : qQ \cdot Qq';$$
or
$$PO \cdot OP' : QO \cdot OQ' = pP \cdot Pp' : qQ \cdot Qq'$$
= ratio of squares of parallel semi-diameters. [Prop. 23.

HYPERBOLA.

PROPOSITIONS PECULIAR TO THE RECTANGULAR HYPERBOLA.

1. $CS^2 = 2CA^2$, $\quad CS = 2CX$, $\quad e = \sqrt{2}$.

2. $PN^2 = AN \cdot NA'$.

3. Latus Rectum $= AA'$.

4. $\qquad\qquad\qquad CN = NG$.

5. *A circle, whose centre is any point* P *on the curve and radius* PC, *intersects the normal on the axes, and the tangent on the asymptotes*
$$PC = PG = Pg = Pr = Pr'.$$

6. *Conjugate diameters are equal, and the asymptotes bisect the angles between them.*

7. *Conjugate diameters are inclined to either axis at angles which are complementary.*

8. *Diameters at right angles to one another are equal.*

9. *The angle between any two diameters is equal to the angle between their conjugates.*

10. *The angles subtended by any chord at the extremities of a diameter* PP' *are equal or supplementary.*

11. *If* CZ *be drawn perpendicular to the tangent at* P,
$$CZ \cdot CP = CA^2.$$

12. *If a rectangular hyperbola circumscribe a triangle it passes through the orthocentre.*

13. *If a rectangular hyperbola circumscribe a triangle, the locus of its centre is the nine-point circle.*

CYLINDER AND CONE.

If a rectangle revolves round one of its sides, the opposite side traces out a surface, called a right circular *cylinder*.

The length of the rectangle may be considered to be indefinitely extended.

The fixed side, about which the rectangle revolves, is called the *axis* of the cylinder.

DEF. A *right circular cylinder* is a surface traced out by a straight line, which moves round the circumference of a circle, and remains always parallel to a fixed straight line, drawn through the centre of the circle, perpendicular to its plane.

DEF. The fixed straight line is called the *axis* of the cylinder.

NOTE. The section of a cylinder by a plane parallel to the axis is two generating lines of the cylinder.

The section of a cylinder by a plane perpendicular to the axis is a circle.

DEF. When a cylinder is cut by a plane, the plane passing through the axis of the cylinder and perpendicular to the cutting plane is called the *axial plane*.

NOTE. The intersection of the axial plane with the cutting plane is an axis of the curve of section; and its intersection with the cylinder is two generating lines.

DEF. A sphere inscribed in a cylinder, so as to touch the cylinder in a circle and the cutting plane at a point, is called a *focal sphere*.

Proposition I.

The section of a right circular cylinder, by a plane inclined to the axis, is an ellipse.

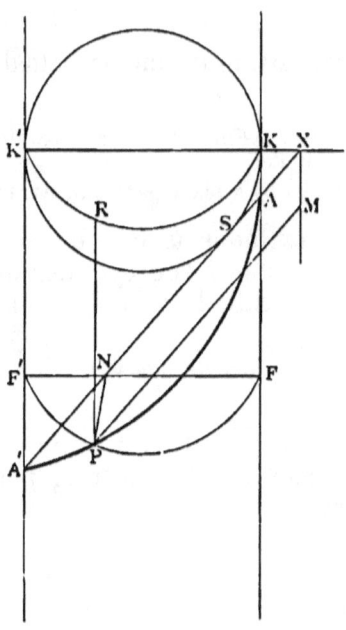

Let APA' be the curve of section. Take the axial plane for the plane of the paper, and let it meet the cutting plane in the straight line $A'AX$ and the cylinder in the generating lines KAF, $K'F'A'$

Draw a focal sphere, touching the cylinder in the circle KRK' and the cutting plane at S.

Let the planes $K'RK$, $A'PA$ meet in the straight line XM.

Through any point P in the curve APA' draw a plane $F''PFN$ perpendicular to the axis of the cylinder, meeting the cutting plane in the straight line PN, the axial plane in the straight line FNF'', and the cylinder in the circle FPF''.

Through P draw the generating line PR, touching the focal sphere at R; also draw PM parallel to NX.

Suppose SP to be joined.

Because the planes APA', FPF'' are both perpendicular to the axial plane, PN is perpendicular to axial plane (Euc. XI. 19); hence PN is perpendicular to both AA' and FF''.

Tangents to a sphere from the same point are equal (Euc. III. 36);

$$\therefore SP = PR = FK,$$

and $\qquad SA = AK$ and $PM = NX$.

But $\qquad FK : NX = AK : AX;$ [Euc. VI. 2.

$$\therefore SP : PM = SA : AX.$$

Now AK is less than AX (Euc. I. 19), therefore $SA : AX$ is a constant ratio less than unity, and APA' is an ellipse whose focus is S and directrix XM.

Proposition I. (Second Method.)

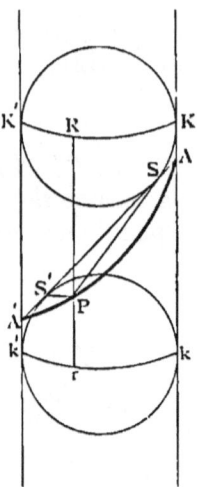

Let APA' be the curve of section. Take the axial plane for the plane of the paper, and let it meet the cutting plane in the straight line AA' and the cylinder in the generating lines KAk, $K'A'k'$.

Draw the two focal spheres touching the cylinder in the circles KRK', krk', and the cutting plane at S and S'.

Through any point P on the curve APA' draw a generating line RPr, touching the focal spheres at R, r. Join PS, PS' which will also touch the focal spheres.

Then $SP = PR$, because they are tangents to a sphere; and $S'P = Pr$.

$$\therefore SP + S'P = PR + Pr = Rr = Kk.$$

Hence the curve is an ellipse whose foci are S, S' and major axis equal to Kk. (Ellipse, 8.)

Proposition I. (Third Method.)

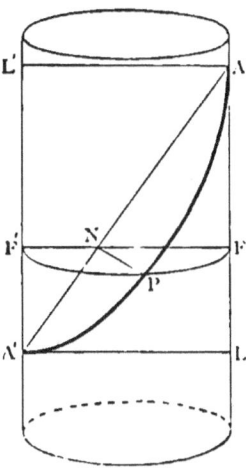

Let APA' be the curve of section.

Take the axial plane for the plane of the paper, let it meet the cutting plane in the straight line AA', and the cylinder in the generating lines AFL, $A'F'L'$.

Through any point P in the curve of section draw a plane $F'PFN$ perpendicular to the axis of the cylinder, meeting the cutting plane in the straight line PN, the axial plane in the straight line FNF', and the cylinder in the circle FPF'.

Draw AL', $A'L$ parallel to KK'.

Because the planes KNK', APA' are both perpendicular to the axial plane, PN is perpendicular to the axial plane (Euc. IX. 19), hence PN is perpendicular to both FF' and AA'.

By similar triangles,
$$AN : NF = AA' : A'L,$$
and $$A'N : NF' = A'A : AL';$$
$$\therefore AN . A'N : NF . NF' = AA'^2 : A'L . AL';$$
$$\therefore AN . NA' : PN^2 = AA'^2 : AL'^2. \quad [\text{Euc. III. 35.}]$$

Hence the section is an ellipse of which AA' is the major axis, and the minor axis is equal to AL'. (Ellipse, 3.)

If a right-angled triangle revolves round one side containing the right angle, the hypothenuse traces out a surface called a right circular *cone*.

The length of the hypothenuse may be supposed to be indefinitely extended in both directions.

The fixed side, about which the triangle revolves, is called the *axis* of the cone.

The angle of the triangle at which the hypothenuse and the fixed side intersect is the *vertex* of the cone.

The complete cone when the hypothenuse is indefinitely extended in both directions consists of two equal and similar sheets on opposite sides of the vertex.

DEF. A *right circular cone* is a surface traced out by a straight line, which moves round the circumference of a circle, and passes always through a fixed point in a fixed straight line drawn through the centre of the circle, perpendicular to its plane.

DEF. The fixed straight line is called the *axis* of the cone.

DEF. The fixed point in the axis is called the *vertex* of the cone.

NOTE. The section of a cone by a plane passing through the vertex is either a point, or two generating lines of the cone.

The section of a cone by a plane, perpendicular to the axis, not through the vertex, is a circle.

DEF. When a cone is cut by a plane, the plane passing through the axis of the cone and perpendicular to the cutting plane is called the *axial plane*.

NOTE. The intersection of the axial plane with the cutting plane is an axis of the curve of section: and its intersection with the cone is two generating lines.

DEF. A sphere inscribed in a cone, so as to touch it in a circle, and the cutting plane at a point, is called a *focal sphere*.

PROPOSITION II.

The section of a cone by a plane not passing through the vertex and not perpendicular to the axis satisfies the definition of a conic section ($SP = e \cdot PM$).

CYLINDER AND CONE.

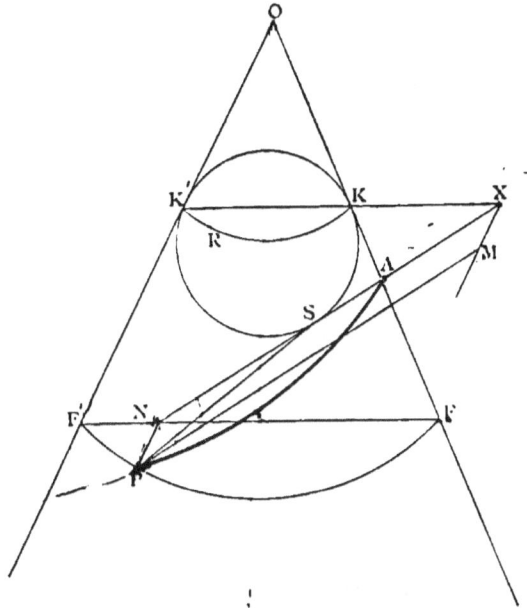

Let AP be the curve of section. Take the axial plane for the plane of the paper, and let it meet the cutting plane in the straight line NAX and the cone in the generating lines $OKAF$, $OK'F'$.

Draw a focal sphere touching the cone in the circle KRK' and the cutting plane at S.

Let the planes $K'RK$, PA intersect in the straight line XM.

Through any point P in the curve AP draw a plane $F''PFN$ perpendicular to the axis of the cone, meeting the cutting plane in the straight line PN, the axial plane in the straight line FNF', and the cone in the circle FPF'.

Suppose the generating line PRO to be drawn, touching the focal sphere at R; also draw PM parallel to NX.

Because the planes AP, FPF' are both perpendicular to the axial plane, PN is perpendicular to the axial plane (Euc. XI. 19); hence PN is perpendicular to both AN and FF'.

Tangents to a sphere from the same point are equal (Euc. III. 36).

Therefore $SP = PR = FK$, and $SA = AK$, and $PM = NX$.

But $FK : NX = AK : AX$;   [Euc. VI. 2.

$\therefore SP : PM = SA : AX$.

Hence APA' is a conic section, having S for focus and XM for directrix.

Proposition III.

A plane section of a cone is an ellipse if its focal axis meets both generating lines in the axial plane on the same sheet of the cone; it is a parabola if its focal axis is parallel to one of these two generating lines; it is a hyperbola if its focal axis meets both these generating lines but on different sheets of the cone.

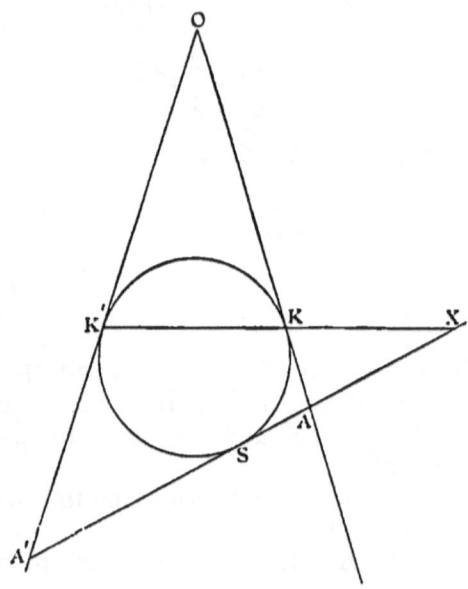

Let the axial plane meet cutting plane in AX, the focal sphere in the circle $KK'S$, and the cone in the generating lines OKA, OK'. Produce $K'K$ and SA to meet in X the foot of the directrix.

Case 1. Produce AS to meet OK' in A'.

angle $OK'X$ > angle $K'XA'$.   [Euc. I. 16.
But angle $OK'X$ = angle OKK' [Euc. I. 5.
     = angle AKX;   [Euc. I. 15.
∴ angle AKX > angle $K'XA'$ or KXA,
∴ $AK < AX$,   [Euc. I. 19.
∴ $SA < AX$,   [Euc. III. 36.

and the curve is an ellipse.

Case 2. If AS is parallel to OK',

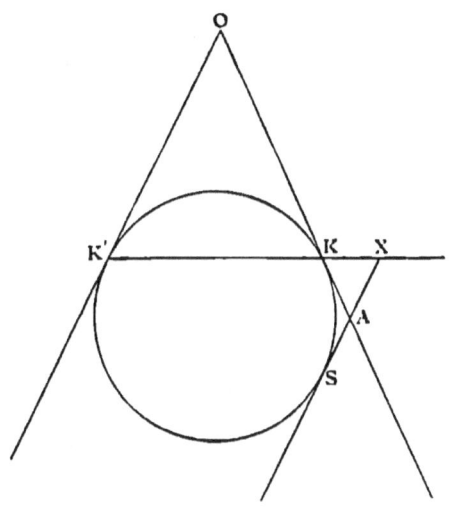

$$\text{angle } AKX = \text{angle } OKK'$$
$$= \text{angle } OK'K$$
$$= \text{angle } KXA; \qquad \text{[Euc. I. 29.}$$
$$\therefore AK = AX, \qquad \text{[Euc. I. 5.}$$
$$\therefore SA = AX, \qquad \text{[Euc. III. 36.}$$

and the curve is a parabola.

Case 3. Produce SA to meet $K'O$ produced in A'.

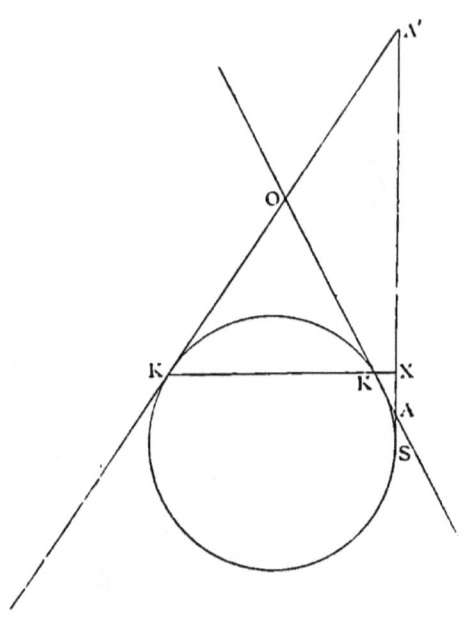

	angle $OK'X <$ angle $K'XA$.	[Euc. I. 16.
But	angle $OK'X =$ angle OKK'	[Euc. I. 5.
	$=$ angle AKX;	[Euc. I. 15.

$\therefore$ angle $AKX <$ angle $K'XA$ or KXA,

$\therefore AK > AX$, [Euc. I. 19.

$\therefore SA > AX$, [Euc. III. 36.

and the curve is a hyperbola.

Proposition IV.

In an elliptic section of a cone the major axis is equal to the distance between the focal spheres measured along a generating line of the cone.

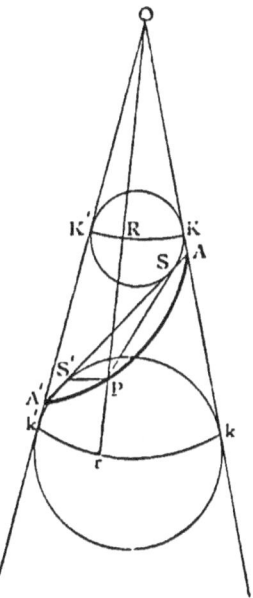

Let APA' be the curve of section. Take the axial plane for the plane of the paper and let it meet the cutting plane in the straight line AA', and the cone in the generating lines KAk, $K'A'k'$.

Draw the two focal spheres touching the cone in the circles KRK', krk', and the cutting plane at S and S'.

Through any point P on the curve APA' draw a generating line RPr, touching the focal spheres at R, r.

Join PS, PS', which will also touch the focal spheres.

Then $SP = PR$, because they are tangents to a sphere; and $S'P = Pr$.

$$\therefore SP + S'P = PR + Pr = Rr = Kk.$$

Hence the curve is an ellipse whose foci are S, S', and its major axis is equal to Kk. (Ellipse, 8.)

132   CYLINDER AND CONE.

Proposition V.

In a hyperbolic section of a cone, the transverse axis is equal to the distance between the focal spheres, measured along a generating line of the cone.

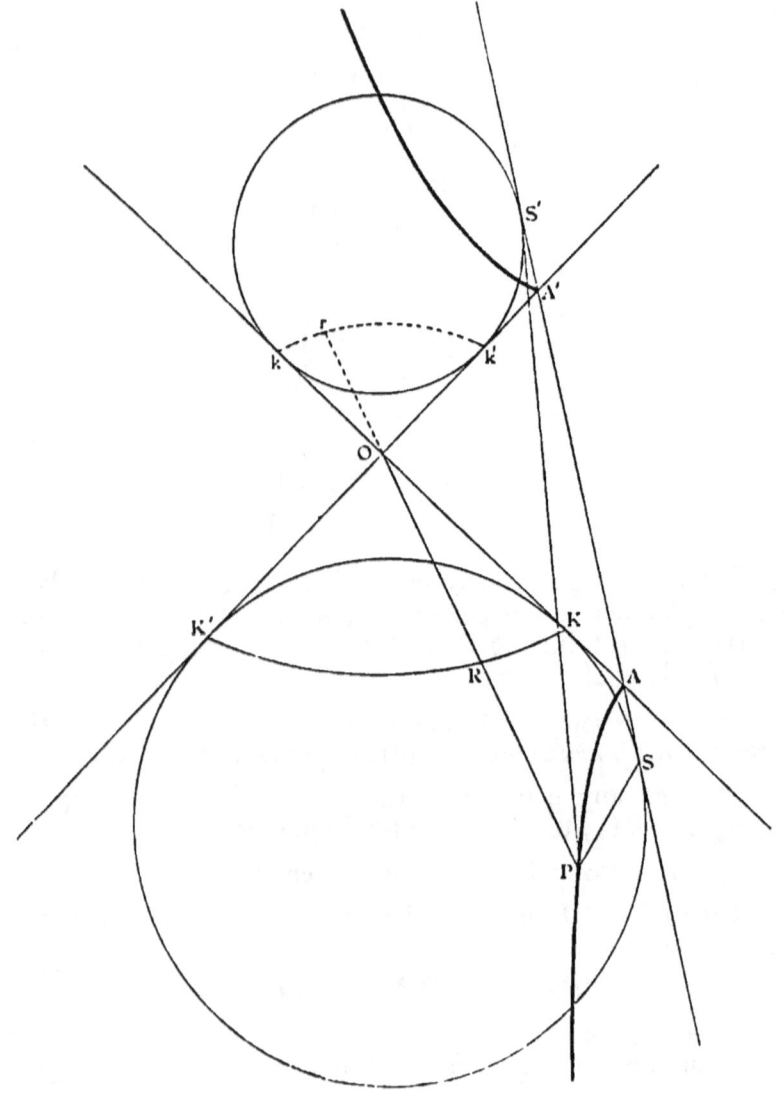

CYLINDER AND CONE.

Let APA' be the curve of section.

Take the axial plane for the plane of the paper and let it meet the cutting plane in the straight line AA', and the cone in the generating lines KAk, $K'A'k'$.

Draw the two focal spheres touching the cone in the circles KRK', krk', and the cutting plane at S and S'.

Through any point P on the curve APA' draw a generating line RPr, touching the focal spheres at R, r.

Join PS, PS', which will also touch the focal spheres.

Then $SP = PR$, because they are tangents to a sphere, and $S'P = Pr$.

$$\therefore S'P \sim SP = Pr \sim PR = Rr = Kk.$$

Hence the curve is a hyperbola, whose foci are S and S', and its transverse axis is equal to Kk. (Hyperbola, 7.)

PROPS. IV. AND V.

The auxiliary circle lies on the surface of the sphere, whose diameter is the line joining the centres of the focal spheres.

Proposition VI.

In a parabolic section of a cone, the latus rectum is a third proportional to the distance of the vertex of the cone from the vertex of the parabola, and the diameter of the circular section of the cone through the vertex of the parabola.

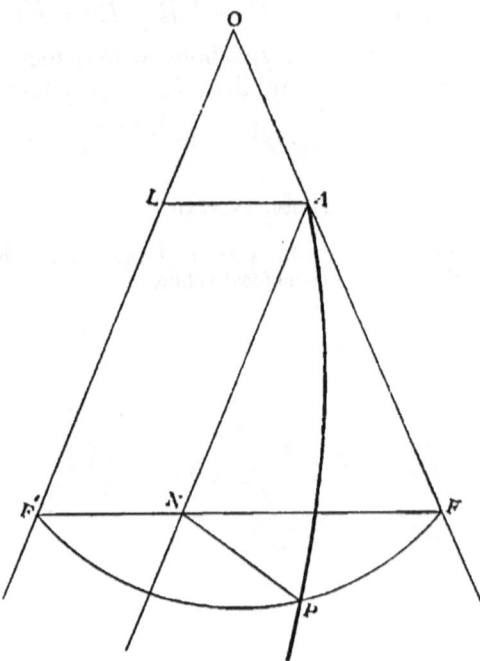

Let AP be the curve of section.

Take the axial plane for the plane of the paper, let it meet the cutting plane in the straight line AN, and the cone in the generating lines OAF, OLF'.

Through any point P on the curve of section draw a plane $F''PFN$ perpendicular to the axis of the cone, meeting the cutting plane in the straight line PN and the axial plane in the straight line FNF'' and the cone in the circle FPF''.

Draw AL parallel to FF''.

Because the planes FPF'', APN are both perpendicular to the axial plane, PN is perpendicular to the axial plane (Euc. XI. 19), hence PN is perpendicular to both FF'' and AN.

Take $4AS$ a third proportional to OL, LA.

By similar triangles
$$AN : NF = OL : LA$$
$$= LA : 4AS;$$
$$\therefore 4AS . AN = NF . LA$$
$$= NF . NF''$$
$$= PN^2.$$

Hence the curve AP is a parabola, of which the latus rectum is $4AS$. (Parabola, 3.)

And $4AS$ is a third proportional to OL, LA.

Proposition VII.

In an elliptic section of a cone, the minor axis is a mean proportional between the diameters of the circular sections of the cone passing through the ends of the major axis.

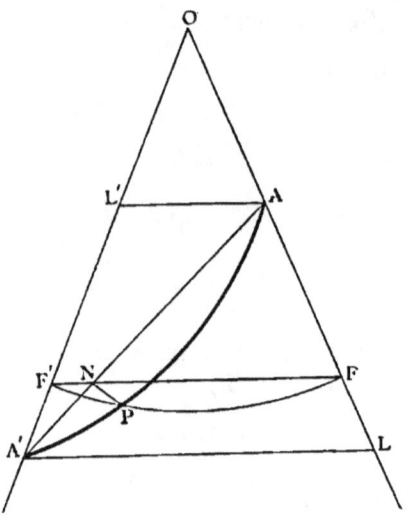

Let APA' be the curve of section.

Take the axial plane for the plane of the paper, let it meet the cutting plane in the straight line AA', and the cone in the generating lines $OAFL$, $OA'F'L'$.

Through any point P on the curve of section draw a plane $F''PFN$ perpendicular to the axis of the cone, meeting the cutting plane in the straight line PN and the axial plane in the straight line FNF'' and the cone in the circle FPF''.

Draw AL', $A'L$ parallel to FF''.

Because the planes FPF', APA' are both perpendicular to the axial plane, PN is perpendicular to the axial plane (Euc. XI. 19), hence PN is perpendicular to both FF'' and AA'.

By similar triangles
$$AN : NF = AA' : A'L,$$
and
$$A'N : NF'' = AA' : AL';$$
$$\therefore AN . A'N : NF . NF'' = AA'^2 : A'L . AL';$$
$$\therefore AN . NA' : PN^2 = AA'^2 : A'L . AL'$$
[Euc. III. 35.

Hence the section is an ellipse of which AA' is the major axis, and the minor axis is a mean proportional between AL' and $A'L$. (Ellipse, 3.)

Proposition VIII.

In a hyperbolic section of a cone, the conjugate axis is a mean proportional between the diameters of the circular sections of the cone, passing through the vertices of the hyperbola.

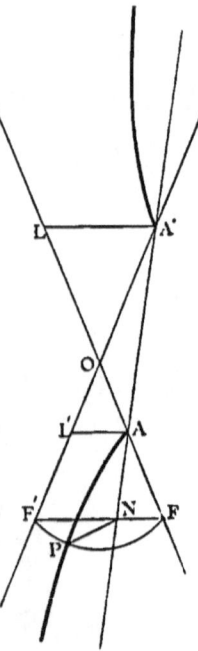

Let AP be one branch of the curve of section, and A' the vertex of the other branch.

Take the axial plane for the plane of the paper, let it meet the cutting plane in the straight line AA' and the cone in the generating lines $LOAF$, $A'OL'F'$.

Through any point P on the curve of section draw a plane $F''PFN$ perpendicular to the axis of the cone, meeting the cutting plane in the straight line PN, and the axial plane FNF'' and the cone in the circle FPF''.

Draw AL', $A'L$ parallel to FF''.

Because the planes FNF'', APA' are both perpendicular to the axial plane, PN is perpendicular to the axial plane (Euc. XI. 19), hence PN is perpendicular to both FF'' and AA'.

By similar triangles

$$AN : NF = AA' : A'L,$$

and $$A'N : NF'' = AA' : AL';$$

$$\therefore AN \cdot A'N : NF \cdot NF'' = AA'^2 : A'L \cdot AL';$$

$$\therefore AN \cdot A'N : PN^2 = AA'^2 : A'L \cdot AL'$$
[Euc. III. 35.

Hence the section is a hyperbola, of which AA' is the transverse axis, and the conjugate axis is a mean proportional between AL' and $A'L$. (Hyperbola, 3.)

Proposition IX.

The asymptotes of a hyperbolic section of a cone are parallel to the two generating lines, which lie in a parallel plane through the vertex of the cone.

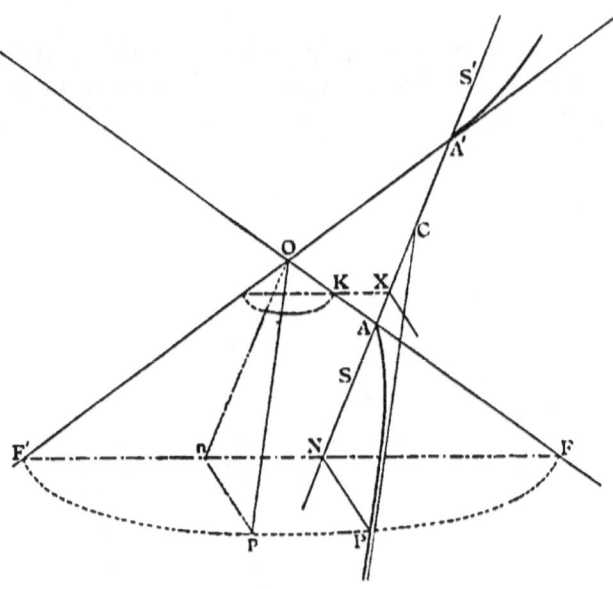

Take the axial plane for the plane of the paper.

Let P be any point on the hyperbola, PN an ordinate, S, S'' its foci, A, A' its vertices, C the centre, and X the foot of the directrix corresponding to the focus S.

Let OF, OF' be generating lines in the axial plane, and $FPF'N$ a plane perpendicular to the axis.

Let the focal sphere touch OF at K, then KX is parallel to FF' (Prop. 2),

and SA is equal to AK. [Euc. III. 36.

Let Opn be a plane parallel to the cutting plane, meeting the cone in a generating line Op, the axial plane in On, the plane FPF' in pn.

The triangles OnF, AXK are similar because On is parallel to AX, and nF to XK.

$$\therefore On : OF = AX : AK$$
$$= AX : AS,$$
$$\therefore OF = e . On ;$$

but the generating lines OF, Op are equal,

$$\therefore Op = e . On.$$

In the figure of Hyperbola, proposition 4,

$$CR^2 = CA^2 + AB^2$$
$$= CA^2 + CB^2$$
$$= CS^2 ;$$
$$\therefore CR = CS = e . CA ;$$

hence pOn is half angle between asymptotes (Hyperbola, 4), but On is parallel to the transverse axis; therefore Op is parallel to an asymptote.

Proposition X.

If through any point two straight lines be drawn, parallel to two fixed straight lines, to intersect a given cone, the ratio of the rectangles contained by the segments of the lines is constant for all positions of the point.

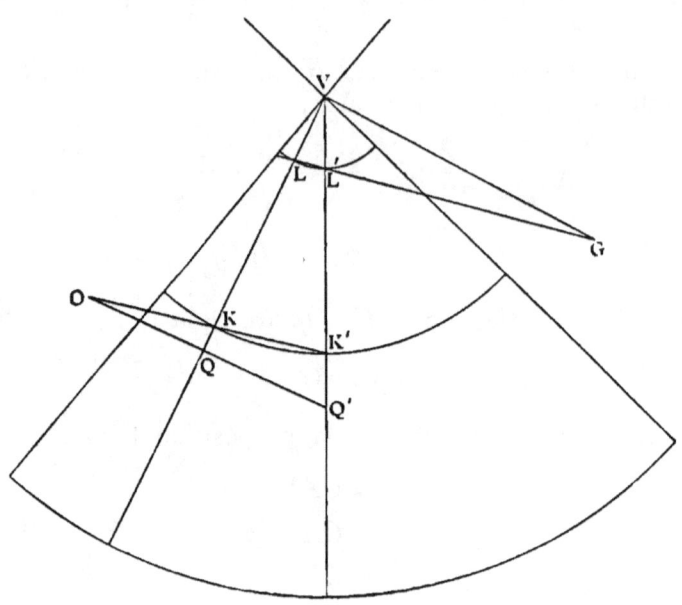

Let OQQ', ORR' be the two lines drawn through O parallel to the two fixed straight lines to meet the cone at QQ', RR'.

Through the vertex V draw VG, VH, parallel to the fixed straight lines; meeting a fixed plane, perpendicular to the axis of the cone at G and H.

ORR' and VH are not shown on the figure.

First consider only the rectangle $OQ \cdot OQ'$.

Let the fixed plane through G and H meet the plane VQQ' in the straight line $GL'L$, and the cone in the circle LL'.

Again let a plane through O, parallel to the fixed plane GH, meet the plane VQQ' in OKK', and the cone in the circle KK'.

The triangles OKQ, GLV lie in one plane and their sides are parallel;
$$\therefore OQ : OK = GV : GL.$$
Similarly $\qquad OQ' : OK' = GV : GL';$
$$\therefore OQ \cdot OQ' : OK \cdot OK' = GV^2 : GL \cdot GL'.$$

Now for all positions of O, GV is constant and the rectangle $GL \cdot GL'$ is constant $\qquad$ [Euc. III. 36.
$$\therefore OQ \cdot OQ' = \lambda \times OK \cdot OK'.$$
Similarly $\qquad OR \cdot OR' = \mu \times OM \cdot OM',$

where λ and μ are constant, and M, M' are the intersections of VR, VR' with the circle KK'.
$$\therefore OK \cdot OK' = OM \cdot OM' \qquad \text{[Euc. III. 36.}$$
$$\therefore OQ \cdot OQ' : OR \cdot OR' = \lambda : \mu.$$

Important propositions to be proved by the reader.

PARABOLA.

1. If POp be a chord of a parabola meeting the axis in O, and PN, pn ordinates, prove that $AN \cdot An = AO^2$. (See Prop. 3.)

2. The circle circumscribing the triangle formed by three tangents to a parabola passes through the focus. (See Prop. 13.)

3. If OQ, OQ' are tangents, and OV a diameter, prove that the angle SOV is equal to the angle $Q'OS$. (See Props. 7, 13.)

4. If P is the end the diameter which bisects a chord QQ', and R the end of another diameter meeting QQ' in M, prove that
$$QM \cdot MQ' = 4SP \cdot RM.$$
(See Prop. 16.)

5. If the diameter through any point R on the curve meets a chord QQ', and a tangent QT at M and T, prove that
$$TR : RM = QM : MQ'.$$
(See Props. 16, 17 and Proof of 19.)

6. If OP touches a parabola at P, and OQR meets at QR, and the diameter through P meets the chord QR in U, prove that
$$OU^2 = OQ \cdot OR.$$
(See Prop. 19.)

7. If a circle meets a parabola in four points A, B, C, D, the common chords AB, CD are equally inclined to the axis of the parabola. (See Prop. 19.)

8. If a circle cuts a parabola in four points the sum of the ordinates of these four points is zero. (See Props. 15, 19.)

9. *If the normals at three points P, Q, R meet in a point, the sum of the ordinates of P, Q, R is zero, and the circle circumscribing the triangle PQR passes through the vertex.* (By analytical geometry.)

10. *If OQ, OQ' be two tangents to a parabola the chord QQ' cuts off from the parabola a segment whose area is two-thirds of the triangle OQQ'.* (See Prop. 16.)

CONIC SECTIONS.

1. *No straight line can meet a conic in more than two points.* (Prop. 2.)

2. *If a circle meets a conic in four points, the chord joining any two of those points makes the same angle with the axis as the chord joining the other two points.* (Ellipse 34.)

3. *To find where a straight line parallel to the axis meets a conic whose focus, directrix, and eccentricity are given.*

[*Cons.* Let the line meet directrix in M. With centre X and radius $e \cdot SX$ describe a circle. Join SM meeting this circle in p, X. Draw SP, SP' parallel to Xp, Xp'. PP' are the required points.]

4. *The semi-latus rectum is a Harmonic Mean between the segments of any focal chord*

$$\frac{1}{SP} + \frac{1}{SP'} = \frac{2}{SL}.$$

$SP : SP' = SN : SN'$
$ = NX - SX : SX - N'X$
$ = SP - SL : SL - SP'.$

5. *The product of the segments of a focal chord varies as the length of the chord.*

6. *Rectangles contained by the segments of any two intersecting chords are proportional to the lengths of the parallel focal chords.* (Ellipse 34.)

7. *Tangents to an ellipse or hyperbola at right angles to one another intersect on a fixed circle, called the Director Circle.* (Ellipse 14.)

8. *Prove*
$$PG : CD = CB : CA \text{ and } Pg : CD = CA : CB.$$
(Ellipse 18 and 33.)

9. *Prove*
$$SP \cdot S'P = CD^2 = PG \cdot Pg.$$
(Ellipse 13 and 18.)

10. *If QQ' be a focal chord, parallel to a semi-diameter CD,*
$$QQ' \cdot CA = 2CD^2.$$

11. *If a diameter of a conic meets the directrix in Z, ZS is perpendicular to the chords bisected by the diameter.*
(Ellipse 11 and 25.)

12. *If OQ, OQ' be tangents to a conic and QQ' meets the directrix in K, OSK is a right angle.* (Ellipse 22.)

13. *If the tangent at P meet any pair of conjugate diameters in T and t,*
$$PT \cdot Pt = CD^2. \qquad \text{(Ellipse 28.)}$$

14. *The projection of the normal PG on the focal distance SP is equal to the semi-latus rectum.* (Ellipse 12.)

15. *If OQ, OQ' are a pair of tangents to an ellipse, and a straight line be drawn from O to meet the curve in K, M, and QQ' in L, $OKLM$ is divided harmonically or*
$$\frac{2}{OL} = \frac{1}{OK} + \frac{1}{OM}. \qquad \text{(Projections.)}$$

16. *If CP, CP' be semi-diameters of a conic at right angles to one another, prove that $\dfrac{1}{CP^2} + \dfrac{1}{CP'^2}$ is constant.*
(Director Circle and Ellipse 33.)

17. *If one straight line passes through the pole of a second straight line, prove that the second straight line passes through the pole of the first.* (Projections.)

SECTIONS OF A CYLINDER AND CONE.

1. *At any point of a plane section the tangent makes equal angles with focal distances and the generating line.*

2. *The semi-minor axis of the section is a mean proportional between the radii of the focal spheres.*

3. *For all sections of a cone the latus rectum varies as the perpendicular from the vertex of the cone on the plane of section.*

4. *An ellipse of any eccentricity may be cut from a right circular cylinder, and may be projected orthogonally into a circle.*

PROBLEMS.

PARABOLA.

1. QSq is a focal chord of a parabola drawn parallel to the tangent at P, PG is a normal. Prove $QS \cdot Sq = PG^2$.

2. Two parabolas have a common focus, and their axes in the same direction: a straight line is drawn through the focus cutting them in four points. Shew that the tangents at these points form a rectangle of which one diagonal passes through the focus.

3. Given the directrix of a parabola and two points on the curve, find the focus. Also draw a tangent parallel to the straight line joining the given points.

4. PNQ is a double ordinate of a parabola and APQ an equilateral triangle; prove that $AN = 3$ times the Lat. Rect.

5. In a parabola the external angle between two tangents is half the angle subtended at the focus by their chord of contact.

6. OQ, OQ' are tangents to a parabola, the chord QQ' meets the axis in R, and OM is drawn perpendicular to the axis, prove that $AM = AR$.

7. If the normal PG at any point of a parabola be divided so that $PQ : QG$ is a constant ratio, prove that the locus of Q is a parabola.

8. Two parabolas have a common directrix, prove that their two common tangents are at right angles to one another.

9. The directrix of a parabola is given and also two tangents: find the focus of the parabola, and the points of contact of the tangents.

10. A chord of a parabola is equal to four times the distance of its middle point from the extremity of the diameter bisecting it; prove that the chord passes through the focus.

PROBLEMS.

11. If OP, OP' are tangents to a parabola meeting the tangent at A in Y and Y', and PP' cuts the axis in K, prove that KY, KY' are parallel to the tangents OP, OP'. (This is true for any diameter, and the tangent at its extremity, not only for the axis.)

12. If PY is a tangent at P to a parabola meeting the tangent at the vertex in Y, and a circle on PY as diameter meets the axis in K and K', prove that PK, PK' produced are normals to the curve.

13. Two chords AB, CD of a parabola are produced to meet in O, and points E, F are taken in AB, CD so that $OE^2 = OA \cdot OB$ and $OF^2 = OC \cdot OD$, prove that EF is parallel to the axis.

14. If a parabola touches the three sides of a triangle its directrix passes through the orthocentre.

15. If two parabolas are drawn through four given points on a circle, their axes intersect in the centroid of the four points.

16. POQ is an acute angle whose sides are tangents to an ellipse at the ends of a focal chord PQ; find the two foci.

ELLIPSE.

17. If the diagonals of a quadrilateral circumscribing a conic intersect in a focus, they are at right angles to each other.

18. Shew how to draw a pair of conjugate diameters in an ellipse inclined at a given angle to one another.

19. P and Q are corresponding points on an ellipse and its auxiliary circle, S is a focus; prove that SP = the perpendicular from S on the tangent to the circle at Q.

20. The normal at P on an ellipse cuts the minor axis in g; Pn is the ordinate to that axis. Prove that
$$Cg : Cn = CS^2 : CB^2.$$

150   PROBLEMS.

21. S is a focus of a given conic, and from a fixed point on the axis a perpendicular is drawn to the tangent at any point P on the curve. Prove that the intersection of this perpendicular with SP lies on a fixed circle.

22. Draw a normal from a given point (1) on the axis of a parabola, (2) on the major axis of an ellipse.

23. From any point P on a common tangent to two ellipses, which have a common focus S, tangents are drawn to the ellipses intersecting another common tangent in Q, R. Prove that the angle QSR is constant.

24. Given an arc of a conic, shew how to determine whether it is part of a parabola, ellipse or hyperbola.

25. Given two tangents to an ellipse and one focus, find the locus of the centre.

26. A tangent is drawn to a conic meeting the directrices in L, M. If S, H be the foci, and LS, MH intersect in N, shew that $LN = MN$.

27. PQ is a double ordinate of a conic, and the straight line joining P to the foot of the directrix cuts the curve in R. Shew that QR passes through the focus.

28. Two chords AP, BQ in an ellipse are produced to meet each other in O; QC, PD are chords parallel to them crossing each other in R, shew that the triangles AOB, CRD are similar, and AB is parallel to CD.

29. If two conics have a common focus and are so placed that they intersect in two points only, then their common chord passes through the point of intersection of the corresponding directrices.

30. A system of parallelograms is inscribed in an ellipse, with their sides parallel to the equi-conjugate diameters: prove that the sum of the squares on its sides is constant.

31. Prove the following construction for drawing a normal to a conic. Draw the ordinate PN, on the axis mark off NK, NL each equal to NP, produce PK, PL to meet the curve again in Q, Q', bisect QQ' in V, then PV is the normal at P.

PROBLEMS.   151

32. An ellipse is inscribed in a quadrilateral $ABCD$, and S is a focus of the ellipse; shew that the angles ASB and CSD are together equal to BSC and DSA.

33. The perpendiculars from the foci on the normal at any point of an ellipse are to one another as the perpendiculars from the foci on the tangent at that point.

34. Given two tangents to a conic and its centre: prove that the locus of its foci is a rectangular hyperbola.

35. If PN, the ordinate at the point P of an ellipse, be produced to meet the tangent at the extremity of the latus rectum in Q, prove that $QN = SP$.

36. An elliptic section of a right cone is projected upon a plane perpendicular to the axis, prove that the focus of the curve of projection is at the point where the axis of the cone meets the plane of projection.

37. If OP, OQ are tangents to an ellipse from a point O on the auxiliary circle, and PCP' a diameter of the ellipse, prove that QP' passes through a focus.

38. In any conic if PQ, PQ' are chords equally inclined to the axis, prove that the circle circumscribing PQQ' touches the conic at P.

39. If two quadrilaterals, inscribed in an ellipse, have three sides of one parallel to three sides of the other, their fourth sides will be parallel. Hence shew how to draw a tangent at any point of an ellipse with a parallel ruler.

(Projections.)

40. If RP is any tangent to a given ellipse at P and SRP a constant angle, prove that the locus of R is a circle.

41. At points Q, Q' on an ellipse OQ, OQ' are tangents, and QG, $Q'G'$ are normals meeting the axis major at G, G', prove that OQG, $OQ'G'$ are similar triangles.

42. Tangents OQ, OQ' subtend equal angles at the foot of the ordinate through O.

43. An ellipse touches a triangle at the middle points of its sides, prove the centre of the ellipse is the centre of gravity of the triangle.   (Projections.)

PARABOLA.

44. If AR, SY are the perpendiculars from the vertex and focus of the parabola on the tangent, prove that
$$SY'^2 = SY . AR + SA^2. \qquad \text{[I. C. S. 1884.}$$

45. P is any point on a parabola, SY is drawn perpendicular to AP meeting the tangent at the vertex in R, prove that AR is one-fourth of PN, the perpendicular from P on the axis. [CLARE, 1888.

46. A parabola touches in A', B', C' the sides of an equilateral triangle ABC, respectively opposite to A, B, C. Prove that AA', BB', CC' meet in the focus of the parabola.
[TRIN. 1887.

47. A parabola rolls on an equal parabola, the vertices originally coinciding; shew that the tangent at the vertex of the rolling parabola always touches a fixed circle. [TRIN. 1887.

48. P, Q are two points on a parabola such that circles described about P, Q as centres and passing through the focus S cut orthogonally in S and R. If the line joining Q to the points of intersection of the circles meet the directrix in T and T', shew that the angle TPT' is equal to half of RPS. [PEMB. 1887.

49. In the parabola if the angle ASP be equal to four-thirds of a right angle, prove that the ordinate at P and the normal at the extremity of the latus rectum intersect on the axis. [MAGD. 1888.

50. Given in position two tangents to a parabola and their points of contact, find the focus and directrix. [QU. 1888.

51. OP, OQ are two tangents to a parabola at P and Q, S is the focus; if OS meet the circle through OPQ again in T, then S bisects OT. [QU. 1888.

52. If PG be the normal at P, prove that the tangent from any point on the parabola to a circle, centre G and radius GP, is equal to the perpendicular from that point on the ordinate of P. [JES. 1888.

53. H is a fixed point on the bisector of the exterior angle A of the triangle ABC; a circle is described upon HA as chord cutting the lines AB, AC in P and Q; prove that PQ envelopes a parabola which has H for focus, and for tangent at the vertex the straight line joining the feet of the perpendiculars from H on AB and AC. [Jes. &c. 1888.

54. Points Y, Y'' are taken on the tangent at the vertex of a parabola so that $SY.SY''$ is constant, and the other tangents through Y and Y'' meet in Q; prove that the locus of Q is a circle. [Joh. 1888.

55. A circle is described touching a parabola at a point P and passing through the focus. If K be the point at which it cuts the axis again, and A the vertex of the parabola, shew that AK is equal to three times the abscissa of P. [Sel. 1888.

56. Two points P, Q are taken on a tangent to a parabola equidistant from the focus. Prove that the other tangents drawn from P, Q will meet on the axis. [Pet. 1886.

57. P, Q, R are points on a parabola, the chord PR intersects the diameter through Q in S. The chord PQ intersects the diameter through R in T. Prove that ST is parallel to the tangent at P. [Clare, 1887.

58. S is the focus and SL the semi-latus rectum of a parabola whose vertex is A. P and Q are any two points in any line through O, the point of intersection of the tangent at A and the diameter through L. Prove that the chord of contact of the tangents from P intersects the chord of contact of the tangents from Q in the straight line which bisects the angle OAS. [Trin. 1886.

59. Prove that, if P be an external point on the axis of a parabola whose focus is S and vertex A, and the tangent at A cut the circle described on PS as diameter in Q, R, then PQ, PR will touch the parabola.

Prove that, if any tangent cut the circle in Q', R', the remaining tangents from Q', R' to the parabola will intersect on the circle. [Trin. 1887.

60. A point moves so that the sum of its distances from a given point and a given straight line is constant, prove that it describes a parabola and find the length of its latus rectum. [Qu. 1887.

61. Give a geometrical construction for the axis of a parabola which passes through the four given points A, B, C, D which are such that AB is parallel to CD. [Jes. 1887.

62. A and P are two fixed points. Parabolas are drawn all having their vertices at A, and all passing through P. Prove that the points of intersection of the tangent at P with the tangent and normal at A lie on two fixed circles, one of which is double of the other. [Joh. 1887.

63. If PN, PL be perpendiculars from P on the axis and the tangent at the vertex, prove that LN always touches a parabola. [Pet. 1886.

64. A variable tangent to a parabola intersects two fixed tangents in the points T and T': shew that the ratio $ST : ST'$ is constant. [Trin. 1886.

65. If QD be drawn perpendicular to the diameter PV of a parabola, then
$$QD^2 : QV^2 = SA : SP.$$
[Trin. 1886.

66. Through Y the foot of the perpendicular from the focus S on the tangent to a parabola at P, YK is drawn parallel to the axis of the parabola, meeting the normal PG in K, SK is joined. Shew that the triangles SKG and SKP are each of them equal to the triangle SPY.
[T. H. 1886.

67. If O be a fixed point, MM' a fixed straight line not passing through O, Q any point in MM', and if on OQ as base an isosceles triangle be described on the side of OQ remote from MM' such that the vertical angle OPQ is always double of the acute angle which OQ makes with MM', shew that the locus of P is a certain parabola. [T. H. 1886.

68. If ABC be a triangle inscribed in a parabola, shew that the sides of ABC are four times as long as those of a triangle formed by the intersection of tangents parallel to them. [I. C. S. 1887.

69. The tangents at P_1, P_2, to the parabola whose vertex is A and axis AN_1N_2 intersect in P, and N_1, N_2 and N are the feet of the ordinates of P_1, P_2 and P. Prove that $P_1N_1 : P_2N_2 :: AN : AN_2 :: AN_1 : AN$.
[I. C. S. 1887.

PROBLEMS.

70. OQ, OQ' are tangents to a parabola, OV a diameter. If OV meet the directrix in K and QQ' meet the axis in N, shew that $OK = SN$; S being the focus.
[I. C. S. 1886.

71. If the tangents at the ends of a focal chord PSQ intersect in D, SD will be a mean proportional between AS and PQ.
[I. C. S. 1883.

72. Find the locus of the centres of circles described within a given segment of a given circle.
[Pet. 1887.

73. PSP', QSQ', RSR' are three chords through the focus S of a given parabola. Prove that the ratio of the areas of the triangles PQR and $P'Q'R'$ is the same as that of the products of the ordinates of P, Q, R and P', Q', R'.
[Pet. 1887.

74. A series of parabolas are drawn to touch two given straight lines, one of them at a given point; shew that the foci lie on a fixed circle and that the directrices pass through a fixed point.
[Trin. 1887.

75. Two equal parabolas, which have a common axis, have their concavities turned in opposite directions. Prove that the locus of the middle point of a chord of either parabola, which is a tangent to the other, is a parabola of one-third the linear dimensions of the given ones.
[Trin. 1887.

76. The normal at P meets the tangent at the vertex in F and the curve again in f. If the axis of the parabola meets at T and G the tangent and normal at P, shew that

$$PF \cdot Pf = TG^2.$$
[T. H. 1888.

77. The normal to a parabola at any point P meets the curve again in Q; T is the pole of the chord PQ, and the line joining T to the focus, S, meets the line drawn through P perpendicular to SP in the point O: prove that $TS = SO$, and that TOQ is a right angle.
[Joh. 1887.

78. V is the middle point of a focal chord QQ' of a parabola, tangents at Q and Q' meet at T; prove that the locus of the intersection of the circle described round the triangle TQQ' and the line TV is a parabola.
[Pet. 1887.

79. From any point on a parabola normals are drawn to the curve at P_1, P_2; shew that the chord P_1P_2 passes through a fixed point. [CLARE, 1887.

80. Two equal similarly situated parabolas have a common axis; a tangent is drawn to one of them meeting the other in P and Q; prove that the perpendicular distance of Q from the diameter through P is constant and that the area of the segment cut off by the chord PQ is constant.
[PEMB. 1886.

81. Determine the point in a parabola at which the normal is equal to a given straight line. [T. H. 1887.

82. If the triangle formed by three tangents to a parabola be isosceles the line joining the intersection of the equal sides to the focus passes through the point of contact of the opposite side with the parabola. [CATH. 1887.

83. Two parabolas having the same focus cut at right angles. Shew that the line joining their vertices passes through the focus and is equal to the focal radius of their point of intersection; also that the locus of the middle points of this line for different pairs of parabolas through the same point is a circle. [JOH. 1886.

84. PQ is a chord of a parabola, PT the tangent at P, and a straight line parallel to the axis cuts the tangent in T, the curve in E, and the chord PQ in F; prove that

$$TE : EF :: PF : FQ.$$ [JOH. 1886.

85. If PN be an ordinate and a chord QNQ' be drawn through N cutting the parabola in Q and Q', then the rectangle contained by the ordinates of Q and Q' is equal to the square on PN. [SEL. 1887.

86. Two fixed straight lines intersect in A, and B is a fixed point; if a circle be described through A and B cutting these lines in C and D, then CD always touches a certain parabola. [SEL. 1887.

87. The normal chord to a parabola at the point whose ordinate is equal to its abscissa subtends a right angle at the focus. [PET. 1885.

PROBLEMS.

88. If a circle passing through the focus of a parabola touches the curve at P and cuts it at L and M, and the axis at N, prove that LP is equal to MN. [CLARE, 1886.

89. Give a geometrical construction for the position of the directrix of a parabola whose axis is parallel to a given line, the parabola passing through two given points and touching a given line through one of them.
[CLARE, 1886.

90. If TP, TQ tangents to a parabola subtend angles at the focus which are constant for all positions of T, prove that the distance between the centres of the circles described about the triangles SPT, STQ will vary as ST^2.
[CLARE, 1886.

91. If PQ be a focal chord of a parabola, and R any point on the diameter through Q: shew that the focal chord parallel to $PR = \dfrac{PR^2}{PQ}$. [TRIN. 1885.

92. Points D, E, F are taken on the sides of a triangle ABC and three confocal parabolas are drawn, one touching BF, FE and EC and the other two the corresponding triads of lines; S is the common focus and the directrices intersect in G, H, K. Prove that the triangles DSG, ESH, FSK are equal to one another. [TRIN. 1885.

93. Two parabolas have a common focus: and from a point T external to both tangents TP, TQ are drawn to one and tangents TR, TS to the other. If the angles PTQ, RTS are supplementary, prove that PR, QS are parallel or meet at the focus. If they are parallel, prove that they are also parallel to the common tangent to the parabolas.
[PEMB. 1885.

94. From two fixed points A, B perpendiculars AP, BQ are let fall on a variable line; prove that the envelope of the line is a parabola when the area of the quadrilateral $ABQP$ is constant. [CAIUS, 1885.

95. The normal at one extremity L of the latus rectum of a parabola meets the curve again in P, the tangent at P cuts the latus rectum produced in M and the axis in T: prove that LM is $\tfrac{4}{3}$ and NT $\tfrac{9}{2}$ times the latus rectum, PN being the perpendicular from P on the axis. [K. 1885.

96. A is the vertex, S the focus and P any point on a parabola; PN is the ordinate at P, and the perpendicular to SP drawn through S meets the normal at P in L; if LM be the ordinate of L, shew that $SM = 2AN$. [Qu. 1886.

97. P, Q are any two points on a parabola, R the middle point of the chord joining them, RM is the ordinate of R drawn perpendicular to the axis and RG drawn perpendicular to PQ meets the axis in G; shew that MG is equal to the semi-latus rectum of the parabola. [Qu. 1886.

98. Prove that the latus rectum is the least focal chord which can be drawn in a parabola. [Cath. 1886.

99. Describe a parabola touching three given straight lines and having its focus in another given line. [Pet. 1861.

100. From S the focus of a parabola a line is drawn parallel to the tangent at a point P meeting the curve in Q; the diameter at P meets SQ in E. Shew that the locus of E is a parabola whose latus rectum is half that of the given one. [Jes. 1861.

101. GR is drawn from the foot of the normal at a point P in a parabola perpendicular to SP cutting the circle described on SP as diameter in L, LS produced meets the tangent at P in O, shew that the ratio of $OS : OP$ is invariable. [Sid. 1861.

102. Parabolas are drawn passing through two fixed points A and B, and having their axes in a given direction; find the locus of the foci. [Joh. 1861.

103. A series of parabolas is described having the same tangent at the vertex as a given parabola, and their foci lying on the given parabola. Shew that they intersect in the focus of the given parabola. [Pet. 1861.

104. The tangent at any point P of a parabola meets a fixed circle whose centre is the focus in Q, R. If the other tangents to the parabola which pass through Q, R meet in T, and if the tangents to the circle at QR meet in U, shew that TU is parallel to the directrix. [Pet. 1882.

105. At the middle point of a focal chord of a parabola a line is drawn perpendicular to the chord and equal to half the chord; find the locus of its extremity. [Clare, 1882.

106. From P, PM is drawn perpendicular to the tangent at the vertex of a parabola and MQ perpendicular to AP; shew that the locus of Q is a circle. [T. H. 1882.

107. Through a fixed point on the axis of a parabola a chord PQ is drawn, and a circle of given radius is described through the feet of the ordinates of P and Q. Shew that the locus of its centre is a circle. [Jes. 1882.

108. If OP, OQ are a pair of tangents to a parabola and PQ cut the axis in R, prove that SR is equal to the distance of O from the directrix. [Jes. 1886.

109. A circle cuts a given circle orthogonally and intersects a given length on a given straight line; shew that the locus of its centre is a parabola, and that the envelope of its chord of intersection with the given circle is a conic.
[Jes. 1886.

110. PSP' is a focal chord of a parabola. The diameters through P, P' meet the normals at P', P in V, V' respectively. Prove that $PVV'P'$ is a parallelogram.
[Jes. 1886.

111. ACP is a sector of a circle, centre C, of which the radius CA is fixed, and a circle is described touching the arc AP externally, and also touching CA and CP both produced; prove that the locus of the centre of this circle is a parabola.
[Joh. 1885.

112. If the direction of the axis of a parabola inscribed in a triangle is given prove the following construction for the focus. Through A one of the angular points of the triangle draw AD, perpendicular to the given direction, cutting the circle in D, through D draw DS perpendicular to the opposite side cutting the circle in S; then S is the focus.
[Pet. 1884.

113. P, Q and R are three points on a parabola whose focus is S. Through R are drawn RU and RV, respectively parallel to the tangents at P and Q, so as to meet the diameter through Q in U and V. Prove *geometrically* that $RU^2 = 4SP \cdot QV$.

Utilize this result to obtain a *geometrical* proof of the following:—

TQ and TR, tangents to a parabola, meet the tangent at P in X and Y. The tangent at the extremity of the diameter through T meets the tangent at P in O. Then if S be the focus, $SP \cdot QR = 2SO \cdot XY$. [Joh. 1886.

114. Two confocal and coaxial parabolas with the concavities in opposite directions are met by any straight line parallel to the axis in P and P' and their common chord QQ' meets PP' in R, shew that $RQ \cdot RQ' : PP'$ is a constant ratio.
[Pet. 1884.

115. The circle circumscribing the triangle formed by three tangents to a parabola passes through the focus: prove that the tangent to this circle at the focus makes with the axis of the parabola an angle equal to the sum of the angles made with the axis by the three tangents to the parabola.
[Pet. 1884.

116. PQ is normal at P to a parabola and T is its pole: shew that PS passes through the vertex of the diameter through T. [Pet. 1885.

117. A straight line moves so that two fixed circles always cut off equal chords from it, shew that it always touches a fixed parabola whose focus bisects the line joining the centres of the two circles. [Pet. 1885.

118. If the ordinate at each point of a parabola be produced below the axis until it is equal to the distance of the point from the focus; prove that the locus of its extremity is another parabola, and that the axes of the curves make with each other an angle equal to half a right angle.
[Clare, 1885.

119. Two fixed tangents to a parabola TQ, TR are met by a variable tangent in X and Y. If a chord of the parabola is drawn parallel to XY and equal to XY, it envelops an equal parabola. [Trin. 1884.

120. A line is drawn through any point P of a parabola perpendicular to the line joining P to the vertex. This line meets the axis in K, and the normal at P meets the axis in G: prove that GK is equal to half the latus rectum.
[Trin. 1884.

121. Through any point on a parabola two chords are drawn equally inclined to the tangent there. Shew that their lengths are proportional to the portions of their diameters intercepted between them and the curve.
[TRIN. 1884.

122. PSp is a focal chord of a parabola, and upon PS and pS as diameters circles are described; prove that the length of either of their common tangents is a mean proportional between AS and Pp. [TRIN. 1885.

123. A straight line PQ cuts two fixed straight lines Ox, Oy which are at right angles, in the points P, Q, and the middle point of PQ lies on a fixed straight line AB. Prove that the straight line PQ is always a tangent to a fixed parabola. [TRIN. 1885.

124. If PG the normal at P meet the axis in G; and if GQ be an ordinate erected from G; prove that the difference between the square on PG and QG is a constant quantity.
[PEMB. 1885.

125. In a central conic if a diameter CT cuts one of its chords QQ' in V, the curve in P and the tangent at Q in T, then $CV \cdot CT = CP^2$; deduce the corresponding proposition for the parabola.

126. If PSQ be a focal chord of a parabola, PG the normal at P, PN the semi-ordinate, and if PN produced meet the diameter passing through Q in H: then HG will be perpendicular to PG. [T. H. 1885.

127. From a point O on the directrix of a parabola are drawn two tangents, and through the focus S two straight lines parallel to these tangents: the part of the directrix intercepted between these parallels will be bisected at O.
[CHR. 1885.

128. An endless string OPQ is fastened at O, and two small beads P, Q slide on it; the string is kept stretched; the beads moving so that OP is always equal to OQ and PQ always fixed in direction: shew that the loci of P and Q are arcs of two parabolas with a common focus at O. [QU. 1885.

129. O is a fixed point on a fixed circle; with any point S on the circle as focus, and the tangent at O as directrix, a parabola is described; shew that the locus of the points of contact of tangents from O to the parabola is a circle.
[Qu. 1885.

130. Given two tangents to a parabola and their points of contact: construct the curve. [Cath. 1885.

131. From any point on a parabola, chords are drawn making equal angles with the tangent at that point; shew that they are to one another as the parallel focal chords.
[Cath. 1885.

132. C is the centre, and D a fixed point on the circumference of a given circle, M is the middle point of any chord RS which is parallel to DC. Prove that CR, CS intersect DM on a certain parabola. [Jes. 1885.

133. The polar of a point O with respect to a parabola meets the axis in U, and a straight line through U at right angles to the polar meets OS in R: prove that $OS = SR$.
[Jes. 1885.

134. Three parabolas have a common tangent. Prove that the points of intersection of their other pairs of common tangents are collinear. [Joh. 1884.

135. If two tangents be drawn to a parabola, the perpendicular from the focus on their chord of contact passes through the middle point of their intercept on the tangent at the vertex. [Joh. 1884.

136. Pairs of equal parabolas are drawn, having a given point S for focus, one touching a given line AB, the other a given line AC. Prove that the envelope of their common tangents is a parabola whose directrix passes through S, and which touches AB and AC at points in one straight line with S. [Joh. 1884.

137. OXP, OYQ, XRY are three tangents to a parabola (focus S) at the points P, Q, R respectively: find the locus of the remaining intersection of the circles SXP, SYP, as the tangent XY varies its position. [Pet. 1883.

138. From the vertex of a parabola lines are drawn parallel to the tangents of the curve: prove that the locus of the points where they meet the corresponding normals is a parabola. [CLARE, 1884.]

139. If two parabolas have a common focus, the line joining it to the intersection of the directrices is perpendicular to the common tangent of the parabolas.
[CLARE, 1884.]

140. Three parabolas are drawn having a common vertex and axis, and their latera recta in geometrical progression: shew that if PQ be the chord of contact of a pair of tangents drawn from a point of the outer to the middle parabola, PQ will touch the inner parabola. [CLARE, 1884.]

141. If any parabola be described touching the sides of a fixed triangle, the chords of contact will pass each through a fixed point. [TRIN. 1884.]

142. A circle round the focus of a parabola as centre cuts the tangent at a point P in the directrix, and also at the point T. TM is drawn perpendicular to SP, produced if necessary. Prove that SM is equal to half the latus rectum.
[PEMB. 1884.]

143. Two tangents OQ, OQ' are drawn from an external point O to a parabola and a perpendicular on the axis from O cuts it in N; prove that NQ, NQ' are equally inclined to the axis. [CAIUS, 1884.]

144. Two parabolas have the same focus and axis, and the tangent at a point P of one parabola meets the tangent at a point Q of the other perpendicularly at T; shew that T is equidistant from the diameters through P and Q.
[CHR. 1884.]

145. A parallelogram circumscribes an ellipse; shew that the circles, each of which passes through the extremities of a side of the parallelogram and through a focus, are all equal. [CHR. 1884.]

146. The portion of the tangent at any point P of a parabola intercepted between the tangents at the extremities of a focal chord subtends a right angle at the point where the diameter through P meets the chord. [CAIUS, 1883.]

11—2

147. A line is drawn through a fixed point, and through the point where a line perpendicular to it meets a fixed line a perpendicular to the fixed line is drawn: prove that the locus of the intersection of this and the first line is a parabola. [CLARE, 1883.

148. Any one of a system of parallel lines cuts two fixed parabolas in P, P' and Q, Q' respectively; through P, P' and through Q, Q' lines are drawn parallel to the axis of the parabola on which they lie; shew that the angular points of the parallelogram so formed are on a fixed conic. [CHR. 1884.

149. A is the vertex of a parabola, P any point on the curve, AP is produced to Q so that $PQ = AP$; and through Q a straight line MQL is drawn perpendicular to AQ meeting the axis in M, if QL be equal to QM shew that the locus of L is a parabola and find the normal at L. [QU. 1884.

150. If the normal at P meet the axis in G the locus of the centre of the circle drawn round APG is a parabola.
[QU. 1884.

151. Having given three tangents to a parabola and the point of contact of one of them, find the focus and draw the parabola. [CATH. 1884.

152. An isosceles triangle is circumscribed to a parabola; prove that the three sides and the three chords of contact intersect the directrix in five points, such that the distance between any two successive points subtends the same angle at the focus. [TRIN. 1886.

153. If PP' be any chord of a parabola perpendicular to the axis and if the diameter through P' meet the tangent and normal at P in Q and R, then will the middle point of QR lie on a fixed parabola. [JES. 1884.

154. The tangents at two points P, Q on a parabola intersect in T and the normals at the same points intersect in O. If TL, ON be drawn at right angles to the axis meeting it in L and N, prove that
$$TL \cdot AL = ON \cdot AS.$$ [JES. 1884.

155. The tangents to a parabola at Q and P intersect in T, and diameters are drawn trisecting PQ. If one of the tangents at their extremities is perpendicular to TP, then will the triangle PTQ be isosceles. [JOH. 1883.

156. If the chord PQ of a parabola be normal at P, and if QP produced meet the directrix in R, prove that the angle RTQ is a right angle. [Joh. 1883.

157. From R, the middle point of PG, the normal to a parabola at P, two other normals RQ, RQ' are drawn to the curve. Prove that QS, $Q'S$ are equally inclined to the axis. [Joh. 1884.

ELLIPSE.

1. The lines AB and AC, at right angles to each other, touch an ellipse whose centre is O, and cut the circle, with centre O and radius OA, a second time in the points B and C respectively. Prove that BC and OA coincide with a pair of conjugate diameters of the ellipse. [I. C. S. 1887.

2. If the normal to an ellipse at a point P meet the axis in G, and PSK be drawn through the focus S to meet the diameter conjugate to CP in K; prove that the ratio of CG to SK will be equal to the eccentricity. [I. C. S. 1885.

3. Construct an ellipse, having given two points as foci, and a given line as tangent. [I. C. S. 1884.

4. Prove that the straight line joining the centre C of an ellipse with the point of intersection of the normals at the ends P, D of a pair of conjugate semi-diameters CP, CD is perpendicular to the straight line PD. [I. C. S. 1885.

5. If X, X' are the feet of the directrices of an ellipse corresponding to the foci S, S', and SY, SY' are the perpendiculars on any tangent, the lines XY, $X'Y'$, will intersect on the axis minor. [I. C. S. 1883.

6. CL is the projection upon the minor axis of the central perpendicular on the tangent to an ellipse at P: prove that if PQ be the diameter of the circle circumscribing the triangle SPS', $PQ \cdot CL = AC^2$. [Pet. 1887.

7. Two normals OA, OB drawn to an ellipse from an internal point O are at right angles. They meet the ellipse again in C and D respectively. Shew that

$$OA : OB :: OC : OD.$$ [Pet. 1887.

8. In an ellipse the perpendicular bisector of a chord P_1P_2 meets the axis major in K, shew that $CK = e^2 CN$, where CN is the abscissa of the middle point of P_1P_2 measured from the centre C, and e is the eccentricity.
[Pet. Pemb. &c. 1888.

9. Lengths CA, CB are taken on two fixed straight lines the sum of whose squares is constant, the parallelogram $ABPC$ is completed: prove that the locus of P is an ellipse making equal intercepts on the lines. [Clare &c. 1888.

10. Any point P on an ellipse is joined to the extremities of two conjugate semi-diameters CA, CB; PA, PB meet CB, CA respectively in B', A'; prove that
$$AA' \cdot BB' = 2CA \cdot CB.$$
[Clare &c. 1888.

11. An ellipse entirely surrounds a concentric circle; shew that the area cut off from the ellipse by tangents to the circle is a maximum or minimum only when the tangent is parallel to an axis of the ellipse, and distinguish the cases.
[Clare &c. 1888.

12. If a parabola can be constructed having its focus at C the centre of an ellipse, and having at P a contact of the second order with the ellipse, shew that
$$3CP^2 = AC^2 + BC^2.$$
If CP be inclined at $45°$ to CA, the axis of the parabola will be inclined at $75°$ to CA. [Clare &c. 1888.

13. If P, Q, R, S be four points on an ellipse such that the centre bisects the parts of an axis intercepted between the chords PQ, RS, then the part of that axis intercepted between the chords PR, QS, and the part between PS, QR will be bisected by the centre. [Trin. 1887.

14. From two points at opposite ends of a diameter of the auxiliary circle, tangents are drawn to the ellipse: shew that the points of intersection lie on the directrices.
[Trin. 1888.

15. A variable right-angled triangle PQR, of which Q is the right angle, is inscribed in a given circle of which the centre is C. If the side QR continually pass through a fixed point S inside the circle, prove that PQ touches an ellipse:

and that if QC and PS intersect in O, the intersection of RO and PQ is the point of contact of PQ with the ellipse.
[LOND. 1st B.A. HON. 1870.

16. Shew that an ellipse has one pair of equi-conjugate diameters. If either extremity of the axis major of an ellipse is joined to an extremity of one of the equal conjugate diameters, the lines drawn from the extremities of the minor axis, parallel to the joining line, will meet the ellipse at the extremities of the other equal conjugate diameter.
[LOND. 1st B.A. HON. 1870.

17. In a given triangle an ellipse is inscribed. If the position of one of the foci is known, shew how to find the ellipse and its points of contact with the sides of the triangle.
[T. H. 1888.

18. If in an ellipse there be inscribed a quadrilateral $PQRS$ such that PQ and SR are parallel, and if tangents to the ellipse be drawn parallel to QR and PS, prove that the straight line joining the points of contact is parallel to PQ and SR. [MAG. 1888.

19. PQ is a chord of a parabola, and T is its pole; an ellipse is drawn with centre on PQ to circumscribe PTQ, K is the pole with regard to the parabola of the tangent at T to the ellipse; prove that TK is parallel to the diameter of the ellipse conjugate to PQ. [K. 1887.

20. P, Q are points in two confocal ellipses, at which the line joining the common foci subtends equal angles; prove that the tangents at P, Q are inclined at an angle which is equal to the angle subtended by PQ at either focus.
[K. 1887.

21. From any point P of a circle PM is drawn perpendicular to the tangent to the circle at a fixed point A on it; shew that the locus of the middle point of PM is an ellipse, and find the centre and axes. [QU. 1888.

22. An ellipse is described having its centre at the focus of a parabola, and having the two diameters of the parabola which pass through the ends of its latus rectum as directrices. Shew that this ellipse will touch the parabola at two points. [QU. 1888.

23. If NP, the ordinate at a point P of an ellipse, produced meet the perpendicular from C on the tangent at P in R, shew that the locus of R is an ellipse, and that the tangents at P, Q, and R to the given ellipse, the auxiliary circle, and the locus of R all meet in a point. [CATH. 1888.

24. Two circles are drawn touching the ellipse at conjugate points P and D respectively and each passing through C: shew that their radii are to one another as CP is to CD. [CATH. 1888.

25. A parabola is described passing through the foci of a given ellipse and having for focus some point on the ellipse. Prove that its directrix always touches the auxiliary circle of the ellipse. Shew also that the point of intersection of the tangents at the foci of the ellipse lies on a circle. [JES. &c. 1888.

26. Through a fixed point O, any chord PQ of a given ellipse is drawn; an ellipse of given magnitude similar and similarly situated to the given ellipse is drawn through P and Q, prove that the locus of its centre is an ellipse. [JES. &c. 1888.

27. An ellipse of given magnitude turns about its centre; prove geometrically that the locus of the pole of any line with respect to it is a circle. [JES. &c. 1888.

28. Of the tangents at the extremities of the minor axis of an ellipse, one meets a latus rectum in E, and the other the corresponding directrix in F; prove that EF is a tangent to the ellipse. [JES. &c. 1888.

29. From P any point on an ellipse a tangent is drawn to the minor auxiliary circle meeting the director circle in Q, R; shew that PQ, PR are equal to the focal distances of P. [JES. &c. 1888.

30. Having given the axes of an ellipse, prove that points on the curve are determined by the following construction. Describe circles on the axes as diameters, and draw a straight line from the centre O meeting the circles in P and Q; the straight line through P parallel to the transverse axis, and the straight line through Q parallel to the conjugate axis, intersect each other in a point R of the ellipse.

Prove also, if a concentric circle be described with radius equal to the sum of the semi-axes, and if the line OPQ meet this circle in V, that VR is the normal to the ellipse at R.
[Joh. 1887.

31. PSQ and $PS'R$ are focal chords of an ellipse; prove that the tangent at P and the chord QR cut the major axis at equal distances from the centre. [Joh. 1888.

32. In the ellipse BC, AC are the semi-minor and semi-major axes and the rectangle $ACBD$ is completed. If the curve bisect SD, where S is the focus, shew that
$$AC^2 + BC^2 = 2AC \cdot CS.$$ [Sel. 1888.

33. The centre of an ellipse, a tangent, the length of the major axis and a point on a directrix are given. Shew how to find the directrices. In what cases will the construction fail? [Pet. 1886.

34. PP' is a diameter of an ellipse, prove that the lines joining the foci to the points where the tangent at P meets the corresponding directrices intersect on the ordinate of P. [Clare, 1887.

35. Two tangents TP and TQ are drawn to an ellipse and any chord TRS is drawn, V being the middle point of the intercepted part; QV meets the ellipse in P'; prove that PP' is parallel to ST. [Trin. 1886.

36. Two points Q and R are taken on an ellipse having DD' for a diameter and QD and RD' meet in P. Prove that an ellipse, similar and similarly situated to the given one, having D for its centre and passing through P, cuts from $D'P$ a chord of which DR is the diameter, and from $D'Q$ a chord of which DQ is the diameter. [Trin. 1886.

37. A tangent at any point P of an ellipse intersects the minor axis in T, and TM is drawn perpendicular to SP produced: shew that the locus of M is a circle.
[T. H. 1887.

38. O is any external point to an ellipse and OS, OS' are drawn to the foci S and S' cutting the curve at the points P and Q, also SQ and $S'P$ are joined intersecting at the point R; a circle is inscribable in the quadrilateral $OPRQ$.
[T. H. 1883.

39. If tangents to an ellipse at points P and P' meet on the auxiliary circle, prove that SP and $S'P'$ are parallel.
[T. H. 1887.

40. If Y and Y' be the feet of the perpendiculars from the foci upon the tangent to an ellipse at P, and PN the ordinate of P, shew that PN bisects the angle YNY'.
[Mag. 1887.

41. If CP, CD be conjugate semi-diameters of an ellipse, PG the normal at P, CZ the perpendicular from C upon the tangent at P, GM the line through G parallel to CD and meeting the straight line drawn from P to either focus in M, shew that PM is a fourth proportional to CB, CD, CZ.
[Mag. 1887.

42. If P and Q be points on an ellipse whose foci are S and H, the four straight lines SP, SQ, HP, HQ, produced if necessary, are tangents to the same circle. [Qu. 1887.

43. The points of contact of tangents to a series of confocal ellipses from a fixed point on either axis lie on a circle. [Qu. 1887.

44. If Y and Z be the feet of the perpendiculars from the foci on the tangent to an ellipse at P, prove that the tangents at Y and Z to the auxiliary circle meet on the ordinate of P, and that the locus of their intersection is an ellipse.
[Cath. 1887.

45. The tangents at the points P, P' of an ellipse meet in T, and the normals meet the axis in G, G' respectively; shew that PG, $P'G'$ subtend equal angles at T.
[Jes. 1887.

46. Prove that the locus of the focus of a parabola which passes through two fixed points, situated on a diameter of a given circle and equidistant from the centre, and which has a tangent to the circle for directrix, is an ellipse whose foci are the two fixed points. [Jes. 1887.

47. Prove that the tangents drawn from the extremity of a diameter of an ellipse to the circle described on the axis minor as diameter form with the focal distances of either extremity of the conjugate diameter a parallelogram the difference of whose sides is equal to the semi-axis major.
[Jes. 1887.

48. Inscribe in an ellipse a triangle similar to a given triangle. [CLARE, 1883.

49. Two conjugate diameters of an ellipse meet the auxiliary circle in P and Q. If P' and Q' be the points on the ellipse corresponding to P and Q, prove that the tangents at P' and Q' are at right angles. [JES. 1887.

50. CA, CB are fixed conjugate diameters and CP, CQ variable conjugate diameters of an ellipse; AP, BQ meet in L; shew that the locus of L is a similar and similarly situated ellipse. [JES. 1887.

51. If TP, TP' be two tangents to an ellipse and PG, $P'G'$ the normals at P and P', and if on TP and TP' points Q, Q' be taken so that $TQ = TG$ and $TQ' = TG'$, shew that $QQ' = 2PU$ when U is the middle point of GG'. [JOH. 1886.

52. If a rectangle circumscribes an ellipse, prove that its diagonals are the directions of conjugate diameters.
[JOH. 1887.

53. TP and PQ are two tangents to an ellipse, one of whose foci is S. PQ and ST intersect in X and from V, the middle point of PQ, a perpendicular VY is drawn to ST; prove that $PV^2 : PX \cdot XQ :: SY : SX$. [JOH. 1887.

54. T, T' lie on CA, CB the semi-axes of an ellipse respectively, and TT' is parallel to AB. Prove that two tangents drawn, one from T, the other from T', to two adjacent quadrants of the ellipse will be parallel to conjugate diameters. [PET. 1885.

55. If SY is the perpendicular from the focus S on the tangent to an ellipse at P, prove that SY, CP meet on the directrix. [PET. 1886.

56. PP' is a diameter of an ellipse, the tangents at P and Q are at right angles: prove that the normal to the ellipse at Q bisects the angle PQP'. [CLARE, 1886.

57. Pp a chord of an ellipse perpendicular to AC is produced to meet the auxiliary circle in P' and p', and the normal at P intersects CP' and Cp' in Q and q: prove that $PQ = Pq = CD$ and $P'Q = BC$. [CLARE, 1886.

58. A tangent to an ellipse at P cuts the major axis in T, and CD is the diameter parallel to PT; prove that
$$TP^2 + CD^2 = ST \cdot TH.$$
[CLARE, 1886.]

59. If P be a point on an ellipse, and the focal distance SP meet the conjugate diameter in E, then the difference of the squares on CP and SE will be constant. [TRIN. 1885.]

60. Two fixed points, Q and R, and a variable point P are taken on an ellipse; prove that the locus of the orthocentre of the triangle PQR is a similar ellipse. [TRIN. 1886.]

61. Two ellipses have a common focus and equal major axes; if one ellipse revolves about its focus in its own plane, prove that its chord of intersection with the other ellipse envelopes a conic confocal with this ellipse. [TRIN. 1886.]

62. From a point R on an ellipse two chords RQ, RQ' are drawn parallel to conjugate diameters CP and CD; the tangent at R meets QQ' produced in T. Prove that
$$\frac{RQ^2}{QT} : \frac{RQ'^2}{QT} = CP^2 : CD^2.$$
[TRIN. 1886.]

63. Two concentric ellipses have the same major axis, and their semi-minor axes are CB and Cb; the ordinate of any point P on the first ellipse meets the second ellipse in p: shew that
$$CP^2 - CB^2 : Cp^2 - Cb^2 = CA^2 - CB^2 : CA^2 - Cb^2.$$
[TRIN. 1886.]

64. A series of ellipses is described with equal major axes. The ellipses have one fixed common focus and one fixed common point. Prove that two consecutive ellipses intersect along the moving focal chord through the fixed point. Also prove that the locus of the point of intersection is an ellipse having the fixed focus and fixed point as foci. [PEMB. 1885.]

65. TP, TQ are tangents to an ellipse at the extremities of conjugate diameters, S is the focus, TR is the perpendicular on SP. Prove that TR is equal to the semi-minor axis. [CAIUS, 1885.]

66. Being given of an ellipse, a focus, a tangent in position, and the length of its minor axis: prove that the locus of its centre is a straight line. [CAIUS, 1885.]

PROBLEMS.   173

67. A given straight line moves with one extremity on the circumference of a circle the radius of which is equal to the given line, and with the other extremity on a fixed diameter of the circle. Shew that every point of the straight line describes an ellipse. Also shew that the sum of the semi-axes of each ellipse is equal to the diameter of the circle.   [T. H. 1886.

68. P is a point on an ellipse, centre C, and P' the corresponding point on the auxiliary circle, CP' meets the normal at P in a point Q: prove geometrically that PQ is equal to the semi-diameter conjugate to CP. [K. 1885.

69. Let PQ be a chord of an ellipse, R the extremity of the diameter CR bisecting PQ, P', Q', R' the corresponding points to P, Q, R on the auxiliary circle; shew that R' is the middle point of the arc $P'Q'$. If CR cut the ellipse in T, and T' be the corresponding point on the auxiliary circle, shew that CT' is perpendicular to PQ. [K. 1885.

70. From a point T on the auxiliary circle of an ellipse an ordinate $TPP'N$ is drawn to the major axis meeting the ellipse in P, the chord of contact of tangents from T in P', and the major axis in N: prove that
$$NP^2 = NP' . NT.$$
[Qu. 1886.

71. A, B are two given points. Ellipses of given eccentricity are drawn so as to pass through A and have AB for normal at A; and so that their axes pass through B: find the loci of the foci. [Cath. 1886.

72. On PN, any ordinate to a fixed diameter of an ellipse, produced if necessary, is taken a point Q, such that NQ is to NP as the diameter conjugate to PN is to the diameter parallel to PN; prove that the locus of Q is an ellipse and determine the positions of the axes. [Pet. 1861.

73. If P, Q be two points on an ellipse such that the sum of their abscissae is constant, the locus of the intersection of the tangents at P and Q is a similar and similarly situated ellipse, passing through the centre of the former. [Caius, 1861.

74. $TYLZ$ is a tangent at L, the extremity of the latus rectum, meeting the axis major in T, and the auxiliary circle in YZ. Shew that the ratio $YL : YZ$ is equal to that of the latus rectum to twice the axis major. [Jes. 1861.

75. If a circle be described upon the major axis of an ellipse, and two diameters be drawn in it at right angles to each other, meeting the circle in Q, q: and if from Q, q, perpendiculars be drawn to the major axis cutting the ellipse in P, p, D, d, respectively, then PCp, DCd are conjugate diameters of the ellipse. [CHR. 1861.

76. TP, TQ are tangents to an ellipse at P, Q; TV, the tangent at T to a confocal ellipse, meets PB produced in V: prove that
$$VP : VQ :: TP : TQ.$$ [TRIN. 1861.

77. If the intercept on the normal to an ellipse made by one of its axes is equal to one of the focal radii vectores to the point whence the normal is drawn, the intercept made by the other axis will be equal to the other focal radius vector. [PET. 1861.

78. From a point P on a parabola a line is drawn perpendicular to the directrix and meeting it in M: prove that the locus of the intersection of AP and SM is an ellipse; A being the vertex of the curve, and S the focus. [CLARE, 1882.

79. Two ellipses have equal minor axes and one focus common. Prove geometrically that the diameters conjugate to the straight lines joining the points of contact of the common tangents in each ellipse are proportional to the major axes. [CLARE, 1882.

80. If S, S' be the foci, P, Q any points on the ellipse; P', Q' the points in which SP, SQ produced are met by the perpendiculars from S' upon the tangent at P and Q respectively; R the intersection of the straight lines PQ, $P'Q'$; then will $S'R$ bisect the exterior angle of the triangle $PS'Q$.

81. From the foci S, H of an ellipse, whose centre is C, SY, HZ are drawn perpendicular to the tangent at P; SP, HZ produced meet in T; TC, YS produced meet in Q, and TS produced meets the circle described about TQY in R. Shew that the locus of R is a circle. [JES. 1882.

82. If from any point P on an ellipse chords PQ, PQ' be drawn parallel to the axes, the normal at P cuts QQ' in a constant ratio. [JES. 1882.

83. From a point T tangents TP, TQ are drawn to an ellipse. If the bisector of the angle PTQ passes through a fixed point O on the major axis of the conic, the locus of T is a circle. [Jes. 1882.

84. If TP, TQ be a pair of tangents to an ellipse from a point T on the auxiliary circle, prove that the quadrilateral formed by joining $SS'PQ$ has two of its sides parallel. Prove also that if O be the intersection of the diagonals the angles CTP, OTQ are equal. [Jes. 1886.

85. The tangents at two points P, Q of an ellipse intersect on a concentric circle. Shew that the straight line PQ touches a concentric and coaxial ellipse whose axes are in the duplicate ratio of the axes of the first ellipse, and shew also that the point of contact of PQ with its envelope never bisects PQ except when PQ is perpendicular to an axis of the two ellipses. [Jes. 1886.

86. P is any point on a fixed circle, PL is drawn in a given direction and is of constant length, and the circle on PL as diameter cuts the given circle again in Q: shew that PQ always touches a fixed ellipse. [Jes. 1886.

87. Prove that any focal chord of an ellipse is a third proportional to the axis major and the diameter parallel to it. [Jes. 1886.

88. PSQ is a focal chord of an ellipse, and the tangents at P and Q meet in Z. Prove that
$$SZ^2 + BC^2 : 2SZ^2 :: CA : PQ.$$ [Jes. 1886.

89. If the normals at conjugate points P and D of an ellipse meet in E, prove that CE is perpendicular to PD. [Joh. 1885.

90. If the circle passing through the foci and one end of the minor axis of an ellipse meet the curve in P and Q, prove that the distances of the tangents at P and Q from the centre are each equal to the distance of a focus from the centre. [Joh. 1885.

91. If a circle roll on the inside of the circumference of a circle of double its radius, prove that any point in the area of the rolling circle traces out an ellipse. Prove that the ellipse traced by the middle point of a radius, and the ellipse

traced by the point on the radius produced, whose distance from the centre of the rolling circle is equal to its diameter, are similar curves. [JOH. 1885.

92. Two parallel tangents to an ellipse touch it at P and Q. Another tangent at R cuts these in T and T'', and PT'' and QT' intersect in V. Prove that RV is parallel to PT and QT'', and is equal to half their harmonic mean.
[JOH. 1885.

93. Prove the existence of the director circle of an ellipse, and prove that the directrix of the ellipse is the radical axis of the director circle and of a point circle at the corresponding focus. [JOH. 1886.

94. If CK be drawn from the centre C perpendicular to the tangent at a point P of an ellipse, and the circle round PKB meet the major axis in M, and with M as centre and CB as radius a circle be described cutting the minor axis in N and N', shew that $MNGN'$ is circumscribable by a circle.
[PET. 1884.

95. An ellipse is drawn through two fixed points A and B and is similar and similarly situated to a fixed ellipse which it cuts in C and D. AC, AD cut the fixed ellipse again in E and F. Shew that the lines CD, EF each pass through a fixed point. [PET. 1884.

96. If S and H be the foci and TP, TQ two tangents to an ellipse at right angles to each other and TM perpendicular to SP; shew that
$$ST \cdot HT = 2TM \cdot AC.$$ [PET. 1884.

97. Two ellipses have the same foci, from points on the outer tangents are drawn to the inner; find the envelope of the chord of contact. [CLARE, 1885.

98. On any chord of an ellipse passing through a fixed point on the major axis, a circle is described having the chord as diameter; prove that the line joining the other two points of intersection of the ellipse and circle passes through a second fixed point on the major axis. [CLARE, 1885.

99. AA' is the major axis of an ellipse of which S and S' are the foci, AR, $A'R'$ are drawn parallel to SP, and $S'P'$ to meet the tangent at P in R and R': prove that
$$AR + A'R' = AA'.$$ [CLARE, 1885.

PROBLEMS. 177

100. If the tangent and normal at a point P of an ellipse meet the major axis in T and G respectively; prove that the circles described on such intercepts as GT have a common radical axis. [CLARE, 1885.

101. Two given ellipses on the same plane have a common focus, and one revolves about the common focus, while the other remains fixed; prove that the locus of the point of intersection of their common tangents is a circle.
[TRIN. 1885.

102. If AQ be drawn from one of the vertices of an ellipse perpendicular to the tangent at any point P, prove that the locus of the point of intersection of PS and QA produced will be a circle, S being one of the foci.
[TRIN. 1885.

103. Through the centre of an ellipse whose foci are S, S' two constant equal lines are drawn parallel to SP, PS' where P is a point on the ellipse: prove that the locus of the fourth angular point of the parallelogram having the equal lines as adjacent sides is a circle. [TRIN. 1885.

104. S and H are foci of an ellipse and T a point on the major axis produced. A circle is described on SH as diameter. Another circle is described to cut the first at right angles and also to cut the major axis at right angles in T. Shew that the latter circle meets the ellipse upon T's polar with respect to the ellipse. [PEMB. 1883.

105. The normal at a point P of an ellipse meets the axes in G, G'. Shew that if CK is the perpendicular from the centre on the tangent at P, O the middle point of CG and O' the middle point of CG', then will $OB = OK = OP$, and $O'A' = O'K = O'P$. [TRIN. 1885.

106. SY and HY' are perpendiculars from the foci S and H of an ellipse upon a tangent and X and X' are the feet of the corresponding directrices; prove that XY and $X'Y'$ intersect on the minor axis. [TRIN. 1885.

107. An ellipse is traced on paper, shew how to find its principal axes. [TRIN. 1885.

108. If P be any point on the tangent at A, the extremity of the major axis of an ellipse, and if PT be the other

C. G. 12

tangent from P to the ellipse, prove that PT is longer than PA. [Pemb. 1885.

109. Two similar and similarly situated ellipses, centres C, C' touch one another at a vertex A: through A is drawn a chord, meeting the ellipses in P, Q respectively: PC, QC'' intersect in R. Find the locus of R. [Pemb. 1884.

110. From any point T on the auxiliary circle of an ellipse tangents are drawn, touching the curve at P and Q. If Pp, Qq be the diameters through these points, shew that Pq, Qp will be focal chords. [Pemb. 1884.

111. The angular points of a triangle are a point on a given ellipse, the centre of the ellipse, and a focus of the ellipse: prove that the locus of the centre of gravity of the triangle is a similar ellipse. [T. H. 1885.

112. If the tangent at any point of an ellipse intersect the tangents at the extremities of the major axis in R and R', then the circle described on RR' as diameter will pass through the foci. [T. H. 1885.

113. Any two fixed points are taken on the major axis of an ellipse; through one a line is drawn parallel to $S'P$, through the other are drawn lines parallel to YS, YS': prove that the latter meet the former in points which are the extremities of a diameter of a fixed circle. [T. H. 1885.

114. PGg normal to the ellipse at P meets the axes in G and g. A circle is described on Gg as diameter and another circle described with P as centre, and cutting the former at right angles, intersects PGg in Q, Q'; prove that the triangles SPQ, $S'PQ'$ are similar. [Chr. 1885.

115. From any point Q of a given circle QR is drawn perpendicularly to a fixed tangent and is divided in P so that $QP : PR$ is in a given ratio; shew that the locus of P is an ellipse. [Qu. 1885.

116. If the diameters through the ends of the latera recta of an ellipse are conjugate diameters, then the line joining the foci subtends a right angle at the ends of the minor axis. [Qu. 1885.

PROBLEMS.

117. If the normal at P of an ellipse pass through the extremity of the minor axis then the circle, described on the line joining the foci as diameter, will touch the tangent at P to the ellipse. [Qu. 1885.

118. A circle is drawn touching an ellipse in two points P and Q symmetrically situated with regard to the axis and passing through the focus S, shew that $SP = SQ =$ latus rectum. [Cath. 1885.

119. Project the following theorem:—If OA and OB be radii of a circle at right angles to each other, and P and Q be points lying respectively on the productions of OA and OB; then PB and QA will meet on the circle if the rectangle $AP \cdot BQ$ be equal to twice the square on the radius of the circle. [Joh. 1884.

120. CA, CB are the semi-axes of an ellipse. If the rectangle $ACBV$ be completed, and the curve bisect SV, shew that $AC^2 + BC^2 = 2AC \cdot CS$. [Pet. 1883.

121. Tangents are drawn to an ellipse from any point on the line through the focus perpendicular to the axis: prove that the length intercepted by them on the corresponding directrix is bisected by the axis. [Pet. 1883.

122. PSQ, PHR are focal chords of an ellipse, QT, RT the tangents at Q and R. Shew that PT is the normal at P. [Pet. 1884.

123. TP, TQ are tangents to an ellipse at P and Q; Cp, Cq are the respective parallel semi-diameters; Tp, PC (produced if necessary) meet in L and Tq, QC in M; PM, QL are produced to meet in V. Prove that TCV is a straight line. [Pet. 1884.

124. A circle and an ellipse have a common diameter, from any point on this diameter tangents are drawn to the ellipse and circle, prove that the lines joining the points of contact are parallel to a fixed line. [Clare, 1884.

125. A series of ellipses have a common centre and have two conjugate diameters given in direction and also the sum of the squares of their axes, prove that they all touch four straight lines. [Clare, 1884.

126. Through the centre of an ellipse whose foci are S, S' two constant equal lines are drawn parallel to SP, PS' where P is any point on the ellipse. Prove that the locus of the fourth angular point of the parallelogram, having the equal lines as adjacent sides, is a circle.
[CLARE, 1884.

127. Through a given point O, a chord OPQ is drawn to a given ellipse: find the stationary values of the rectangle OP . OD, and distinguish between the maximum and minimum values.
[TRIN. 1883.

128. P, Q, R are three points on an ellipse, centre C, RP, RQ meet the diameter ACA' which bisects PQ in N and T. Shew that $CN . CT = CA^2$.
[TRIN. 1884.

129. The diameter parallel to any focal chord of an ellipse is equal to the chord joining the points on the auxiliary circle which correspond to the extremities of the focal chord.
[TRIN. 1884.

130. Shew how to draw a focal chord of given length in a given ellipse and prove that if the two chords so drawn be PQ and $P'Q'$, then a circle can be described round $PP'QQ'$.
[TRIN. 1884.

131. If a triangle can be inscribed in an ellipse with its centre of gravity at the centre of the ellipse the triangle must be the greatest triangle which can be inscribed.
[TRIN. 1884.

132. If the normal PG to an ellipse pass through B, prove that BG is equal to half the distance between the foci.
[PEMB. 1884.

133. If a tangent, its point of contact and one focus of an ellipse be given, find the locus of its centre.
[CAIUS, 1884.

134. On TQ, TQ' a pair of tangents to an ellipse, whose foci are S and H, TR, TR' are taken equal to TS and TH respectively; prove that RR' is equal to the major axis, and that if TS cut RR' in W, TW is equal to TQ.
[CAIUS, 1884.

135. A given straight line moves with one extremity on the circumference of a circle the radius of which is equal to the given line, and with the other extremity on a fixed diameter of the circle. Shew that every point of the straight

PROBLEMS. 181

line describes an ellipse. Also shew that the sum of the semi-axes of each ellipse is equal to the diameter of the circle.
[Mag. 1884.

136. If the tangent at a point P of an ellipse meet the tangent at the vertex A in T and S' be the focus further from A, then TA is equal to the perpendicular from T on $S'P$.
[Qu. 1884.

137. If CY, CZ be drawn perpendicular to the tangents to an ellipse at P and D conjugate points, and D' be the opposite end of the diameter CD, shew that PD' is the diameter of the circle described round the triangle YCZ.
[Qu. 1884.

138. Having given the auxiliary circle of an ellipse and a tangent to the ellipse touching the ellipse at a given point, find the foci of the ellipse. [Cath. 1884.

139. If AA' is the transverse axis of an ellipse, and if Y, Y' are the feet of the perpendiculars let fall from the foci on the tangent at any point of the curve, prove that the locus of the point of intersection of AY and $A'Y'$ is an ellipse.
[Trin. 1885.

140. The perpendicular from C on QQ' meets the auxiliary circle in R; through C a line is drawn parallel to PR meeting a perpendicular to QQ' through V in O. Prove that, if an ellipse be described through Q and Q' with O as centre and major axis equal to that of the given ellipse, it will have its minor axis equal to DCD'. [Trin. 1886.

141. Two tangents TP and TQ are drawn to an ellipse and any chord TRS is drawn, V being the middle point of the intercepted part; QV meets the ellipse in P'; prove that PP' is parallel to ST. [Trin. 1886.

142. Two points Q and R are taken on an ellipse having DD' for a diameter, and QD and RD' meet in P. Prove that an ellipse, similar and similarly situated to the given one, having D for its centre, and passing through P, cuts from $D'P$ a chord of which DR is the diameter, and from $D'Q$ a chord of which DQ is the diameter. [Trin. 1886.

143. Through the foci S, H of an ellipse two lines PSP', QHQ' are drawn meeting two tangents PQ, $P'Q'$ and such

that PP', QQ' are bisected in S and H respectively. Shew that a circle can be described about the quadrilateral $PQQ'P'$. [Jes. 1884.

144. In the ellipse if the perpendiculars from G and C on CP and the tangent at P meet in H, and the circle on CH as diameter meet the tangent at P in L, prove that CL is equal to the tangent drawn from P to the circle described on the axis minor as diameter. [Jes. 1884.

145. The locus of the intersection of tangents to an ellipse at right angles is a circle. [Jes. 1884.
If the tangent at P cut this circle in T, prove that TP subtends at the foci angles which are complementary.

146. A circle passing through the foci of an ellipse intersects the curve at P and Q on opposite sides of the axis. Prove that the sum of the squares of the perpendiculars from the centre on the tangents at P and Q is equal to the square on AC. [Joh. 1883.

147. From the foci S, H, SO, HO' are drawn perpendicular to SP, HP to meet the normal at P in O, O'. Shew that OO' is bisected by the minor axis. [Pet. 1883.

HYPERBOLA.

1. Give in magnitude and position the two axes ACA', BCB' of a hyperbola, construct geometrically a pair of conjugate diameters PCP', DCD', which shall contain a given angle. [I. C. S. 1886.

2. A straight line cuts a pair of conjugate diameters of a hyperbola in P and D, and a second pair in P' and D'; if O be the middle point of the line intercepted between the asymptotes, prove that
$$OP \cdot OD = OP' \cdot OD'.$$ [I. C. S. 1886.

3. Given one focus, a tangent, and the length of the minor axis a hyperbola, shew that the locus of the centre is a straight line. [I. C. S. 1885.

4. If two tangents of a hyperbola intersect on one branch of the conjugate hyperbola, prove that their chord of contact touches the other branch. [I. C. S. 1885.

5. Through N the foot of the ordinate of a point P on a hyperbola draw NQ parallel to AP to meet CP in Q. Prove that AQ is parallel to the tangent at P. [I. C. S. 1884.

6. Two angular points of an equilateral triangle are respectively the centre and one focus of a hyperbola, and one side of the triangle is an asymptote. Find where the other two sides are cut by the curve. [I. C. S. 1883.

7. If two sides of a triangle are fixed in direction and the third passes through a fixed point, the locus of the centres of the circles circumscribing the triangle will be a hyperbola.
[I. C. S. 1883.

8. A circle is described having for diameter a chord of a rectangular hyperbola with its ends on different branches. Prove that the perpendiculars drawn to this chord from the other points of intersection of the circle and hyperbola are tangents to the hyperbola. [Pet. 1887.

9. Given in position the asymptotes and one tangent to a hyperbola, shew how to construct the curve.
[Pet. 1887.

10. A circle and a rectangular hyperbola intersect in four points which lie on a given parabola; prove that an axis of the hyperbola is parallel to the axis of the parabola; and shew that whatever curve the centre of the hyperbola (or circle) describes, the centre of the circle (or hyperbola) will describe an equal curve, the two centres moving over their respective curves in opposite directions. [Pet. 1887.

11. A parabola and rectangular hyperbola, one of whose asymptotes is the axis of the parabola, each circumscribe the triangle PQR whose sides cut the axis of the parabola in p, q, r, respectively. If A be the vertex of the parabola, and PN the ordinate of P, prove that
$$Aq + Ar = AN.$$
[Pet. Pemb. &c. 1888.

12. With each pair of three given points as foci, a hyperbola is drawn passing through the third point: shew that the three hyperbolas thus drawn intersect in a point.
[TRIN. 1888.

13. Shew that all the conics which pass through the three vertices of a triangle and the intersection of its three perpendiculars are equilateral hyperbolas: and determine the locus of the centre of these hyperbolas.
[LOND. 1st B.A. HON. 1872.

14. Two points P, Q are taken on a hyperbola so that the tangent at P and a parallel through Q to one asymptote intersect on the other asymptote; shew that the tangent at Q and a parallel through P to the second asymptote intersect on the first asymptote.
[TRIN. 1888.

15. Given a hyperbola traced on paper, how would you find its transverse and conjugate axes and its asymptotes?
[T. H. 1888.

16. Having given the asymptotes of a hyperbola and a point on the curve, find the foci, directrices, and vertices.
[C. C. C. 1888.

17. C is the centre of a rectangular hyperbola, a straight line LQ is drawn parallel to one asymptote CM meeting the other in L, and the angle QCM is bisected by a straight line which meets the hyperbola in P; shew that CQ is proportional to CP^2, Q being any point on the line LQ.
[CATH. 1888.

18. The perpendiculars drawn from the foci of a rectangular hyperbola on the tangent at any point P meet the curve in points K, L, M and N. Prove that $KLMN$ is a parallelogram two of whose sides are at right angles to the diameter through P.
[JES. &c. 1888.

19. One asymptote and three points of a hyperbola being given, construct the other asymptote.
[JES. &c. 1888.

20. If P be any point of a hyperbola and AA' its transverse axis, and if $A'P$ and AP meet a directrix in E and F, prove that EF subtends a right angle at the corresponding focus.
[JOH. 1888.

PROBLEMS.   185

21. With two sides of a square as asymptotes, and the opposite point as focus, a rectangular hyperbola is described; shew that it bisects the other sides.   [JOH. 1888.

22. An ellipse is drawn having its axes, major and minor, coincident in direction and magnitude with those of a hyperbola: from any point T on either asymptote, tangents TQ, TQ' are drawn to the ellipse: prove that the circle described round TQQ' passes through the centre of the hyperbola.   [CLARE, 1887.

23. $ABCD$ is a rectangle. Two equilateral hyperbolas having their asymptotes parallel to the sides of the rectangle pass through A and C, and B and D, respectively. Prove that the polar of the centre of one hyperbola with respect to the other coincides with the polar of the centre of the latter with respect to the former.   [TRIN. 1886.

24. P is a point in the plane of a triangle ABC, such that the perpendiculars from A, B, C upon PB, PC, PA respectively meet in a point. Shew that the locus of P is a hyperbola circumscribing the triangle ABC and passing through the points of intersection of the perpendiculars let fall from A, B, C upon the opposite sides of the triangle with the straight lines drawn from B, C, A respectively perpendicular to BA, CB, AC.   [TRIN. 1886.

25. Prove that the parallel focal chords of conjugate hyperbolas are to one another as the eccentricities of the hyperbolas.   [TRIN. 1887.

26. Find the locus of the intersection of the tangent with a straight line drawn from the focus making a fixed angle with the tangent.   [TRIN. 1887.

27. P is a point on a hyperbolic branch whose vertex is A, LPL' is the tangent at P terminated by the asymptotes, and $MPAM'$ is a straight line terminated by lines drawn through the further vertex parallel to the asymptotes; shew that LM and $L'M'$ are parallel.   [MAG. 1887.

28. If P and Q be any two points on a rectangular hyperbola, C the intersection of the axes, PT the tangent at P, QM and QN the perpendiculars from Q upon CP and PT respectively, shew that CM and CN are equal.

[MAG. 1887.

29. If P be any point of a hyperbola whose foci are S and H, and if the tangent at P meet an asymptote in T, the angle between that asymptote and HP is double the angle STP. [K. 1886.

30. If a tangent at P meets the asymptotes in L and M the locus of the centre of the circle circumscribing the triangle LCM is a hyperbola having its asymptotes at right angles to the original ones. [Qu. 1887.

31. Ox, Oy are any two fixed straight lines; A lies on Ox and B on Oy and $OA = OB$. Through A, B, any two parallel lines AM, BN are drawn meeting Oy and Ox respectively in M and N; shew that the locus of the middle point of MN is a hyperbola. [Cath. 1887.

32. A circle which passes through two fixed points S, S', cuts two fixed straight lines, which are perpendicular to SS' and equidistant from its middle point, in the points P, Q, and P', Q'. Shew that if PP' be not parallel to SS', it will touch a fixed conic whose foci are S, S'.
[Jes. &c. 1887.

33. A rectangular hyperbola is drawn passing through two fixed points P, Q on a fixed conic, and having an asymptote parallel to a given straight line: shew that if it cuts the given conic again in R and S, the straight lines PR and QS intersect on a fixed conic. [Jes. 1887.

34. OX, OY are fixed straight lines; A is a fixed point on OX and P a variable point on OY; PM is drawn perpendicular to AX and Q taken on PM so that $AQ = PM$; find the locus of Q. [Jes. 1887.

35. P is any point on a circle of which AB is a fixed diameter. Through B a line is drawn to meet AP produced in Q so that BP, BQ make equal angles with AB. Find the locus of Q. [Jes. 1887.

36. If a triangle ABC be inscribed in a rectangular hyperbola, prove that its orthocentre P lies on the hyperbola.
If through P chords PA', PB', PC' be drawn parallel to the sides of the triangle, prove that AA', BB', CC' are parallel. [Joh. 1886.

37. A and C are points on opposite branches of a rectangular hyperbola, and the circle described on AC as diameter meets the curve again in B and D. Prove that the distances of any point on the hyperbola from the sides of the quadrilateral are proportionals. [JOH. 1886.

38. The base AA' of a triangle is fixed in magnitude and position: prove that if the difference of the base angles is a right angle, the locus of the vertex is a rectangular hyperbola.

If PN is the perpendicular on AA' and NQ, NQ' the tangents from N to the circle on AA' as diameter, prove that PQ passes through A' and PQ' through A; and also, if QQ' intersect AA' in M, that PM is the tangent at P.
[JOH. 1887.

39. If a family of rectangular hyperbolas be described about a triangle, their centres will all lie on the nine-point circle.

If the triangle be right-angled, all the hyperbolas will have a common tangent at the right angle. [PET. 1886.

40. Prove geometrically that the locus of points on a system of confocal ellipses where the tangents are parallel to a given line is an equilateral hyperbola. [CLARE, 1886.

41. If the conjugate diameters PCp, DCd of an ellipse be the asymptotes of a hyperbola, QQ' one of the common chords, $Q'R$, QR chords of the ellipse parallel respectively to CD and CP, prove that $Q'R : QR :: CD : CP$. [CLARE, 1886.

42. Prove that the common chords of a hyperbola and circle may be grouped in pairs which meet the asymptotes in concyclic points; and that these circles are all concentric with the original circle. [TRIN. 1886.

43. Having given, in a triangle, its base and the difference of its base angles, prove that the locus of the vertex is a rectangular hyperbola. When is the base of the triangle the transverse axis? [CAIUS, 1885.

44. If two concentric rectangular hyperbolas have a common tangent the angle between their transverse axes will be half the angle between the straight lines from the centre to the points of contact. [T. H. 1886.

45. In a hyperbola, supposing the two asymptotes and one point of the curve to be given in position, find the position of the vertices. [T. H. 1886.

46. Four tangents to a hyperbola form a rectangle. If one side AB of the rectangle cut a directrix of the hyperbola in X and S be the corresponding focus, shew that the triangles XSA, XSB are similar. [Chr. & E. 1885.

47. In the rectangular hyperbola, the angle between a chord PQ and a tangent at P is equal to the angle subtended by the chord PQ at the other extremity of the diameter through P.

48. Two rectangular hyperbolas touch one another in P and intersect in R and S. Prove that the circle on RS as diameter passes through P and the extremities of the two diameters through P. [Chr. & E. 1885.

49. If an equilateral triangle be inscribed in a rectangular hyperbola, find the locus of the centre of its circumscribing circle. [Qu. 1886.

50. In the rectangular hyperbola, prove that the portion of the normal at any point intercepted between the point and the axis, is equal to that semi-diameter of the conjugate hyperbola which is perpendicular to the normal.
[Joh. 1861.

51. Parabolas are drawn passing through two fixed points A and B, and with their axes parallel to a given straight line; if a tangent be drawn at right angles to AB, prove that the locus of its point of contact is a hyperbola.
[Joh. 1861.

52. A straight line moves between two straight lines at right angles to each other so as to subtend a right angle and a half at a fixed point on the bisector of the right angle; prove that it always touches a rectangular hyperbola.
[Joh. 1861.

53. Prove that a rectangular hyperbola, confocal to a given ellipse, intersects it at the extremities of its equi-conjugate diameters. [Pet. 1861.

54. If a parabola be described with any point on a hyperbola for focus, and passing through one of the foci of

the hyperbola, shew that its axis will be parallel to one of the asymptotes. [PET. 1882.

55. The tangent to a parabola at P meets the tangent at the vertex in Y. The ordinate PN is produced to R so that $RN = PY$. Shew that the locus of R is a rectangular hyperbola. [JES. 1882.

56. A and B are fixed points on a given circle, and CD is any chord of given length. If CD be drawn parallel to AB, and if AE, BD meet in O, the locus of O is a rectangular hyperbola. [JES. 1882.

57. Given the auxiliary circle of a hyperbola and a point on the curve, shew that the locus of the foci is an hyperbola. [JES. 1886.

58. Shew that the locus of the intersection of two equal circles which touch two given parallel straight lines at given points A and B and whose centres are on the same side of AB is a hyperbola. [JES. 1886.

59. Shew that the angle between two tangents to a rectangular hyperbola is equal or supplementary to the angle which their chord of contact subtends at the centre, and that the bisectors of these angles meet on the chord of contact.
[JES. 1886.

60. The tangent at a point P of a rectangular hyperbola meets the asymptotes in K and L, and the normal at P meets the axis in G; find the centre of the circle circumscribing the quadrilateral $CKGL$. [JOH. 1885.

61. Two hyperbolas have the same transverse axis and a line perpendicular to it meets them in points P and P'. Prove that the tangents at P and P' meet on the transverse axis. [PET. 1884.

62. A tangent to a hyperbola at a point P meets an asymptote in T. A line $R'PR$ is drawn parallel to this asymptote, to meet a directrix in R' and the line ST in R, where S is the focus corresponding to the directrix; prove that $R'P = RP$. [CLARE, 1885.

63. Shew that if the tangent at a point P of a hyperbola meet an asymptote in T, the angle between CT and HP will be double the angle STP; where C is the centre, and S and H the foci of the curve. [TRIN. 1884.

64. Shew that if CP, CD be conjugate semi-diameters of a hyperbola whose foci are S and H, then the distance of D from a line drawn through C parallel to HP will be equal to the semi-minor axis. [TRIN. 1885.

65. The tangent to a hyperbola at a point P meets the asymptotes in Q, q; QM, qm are the ordinates of Q, q, and CT the perpendicular from the centre on the tangent at P.

If TM, Tm meet the normal at P in K, L respectively, shew that $OKgL$ is a rhombus. [PEMB. 1885.

66. Defining the hyperbola to be the envelope of the line which cuts off from two fixed lines a triangle of constant area, prove that the hyperbola has two asymptotes and that the line touches the curve at its middle point.
[G. & C. 1885.

67. Prove that the angle between the tangents at a point of intersection of two concentric rectangular hyperbolas is double of the angle between their transverse axes.
[T. H. 1885.

68. Let PQ be any diameter of a rectangular hyperbola and let a circle be described with centre P and radius PQ, then if A, B, C be the other points in which the circle cuts the hyperbola, the triangle ABC is equilateral.
[K. 1884.

69. A circle meets a given rectangular hyperbola in A, A', P, P', prove that the tangents to the hyperbola at P, P' intersect in a point lying on the diameter at right angles to AA'. [CHR. 1885.

70. S is the focus of a parabola whose vertex is A, and SA meets the directrix in X; SXH is an angle of $60°$ and SH is perpendicular to SX, shew that a hyperbola may be described with S and H as foci touching the parabola in a point P whose focal distance is equal to the latus rectum.
[QU. 1885.

71. Through a given point P any straight line is drawn meeting two fixed straight lines in P' and Q'; a point Q is taken on $P'PQ'$ so that $QQ' = PP'$; shew that the locus of Q is a hyperbola. [CATH. 1885.

72. The tangent and normal at any point of a hyperbola intersect the asymptotes and axes respectively in four points which lie on a circle passing through the centre of the hyperbola, and the radius of this circle varies inversely as the perpendicular from the centre upon the tangent.
[Joh. 1884.

73. If the asymptotes of a hyperbola be inclined to each other at an angle equal to half a right angle, find (and trace) the locus of the orthocentre of the triangle CHK, where H and K are the points in which lines through P parallel to one asymptote meet the other respectively.
[Pet. 1883.

74. If the tangent at a point L meets an asymptote in T, and the chords joining L to two other points M and N, meet the asymptote in A and O; prove that $TA = A'O$, where A' is the point in which MN meets the asymptote.
[Clare, 1884.

75. $ABCD$ is a parallelogram; from any point E in BC a perpendicular EF is drawn on AD, and EG is drawn at right angles to AE, the points F and G being on AD, on AB a point K is taken so that $AK = FG$, prove that FG always touches a fixed hyperbola.
[Trin. 1884.

76. From any point P in a hyperbola, perpendiculars PM, PN are drawn to the asymptotes, and PN meets the curve again at P', prove that the ratio of PM to $P'N$ is the same for all positions of P.
[Pemb. 1884.

77. Parallel tangents are drawn to a system of circles which pass through two fixed points; shew that the locus of the points of contact is a rectangular hyperbola.
[Chr. 1884.

78. The points A, B, C, D, lie on a hyperbola, and the lines AB, CD intersect on an asymptote; find the other asymptote.
[Pet. 1884.

79. Tangents are drawn to a rectangular hyperbola from a point T on the transverse axis, meeting the tangents at the vertices in Q and Q'. Prove that QQ' touches the auxiliary circle in a point R such that RT bisects the angle QTQ'.
[Trin. 1885.

80. A line is drawn parallel to the side AC of a triangle ABC meeting, in P and Q respectively, AB and the tangent at C to the circle circumscribing the triangle ABC. Shew that the locus of the intersection of CP, BQ is a rectangular hyperbola. [Jes. 1884.

81. Given an asymptote and two points on an hyperbola, shew that the envelope of the axis is a parabola.
[Jes. 1884.

82. Chords of a hyperbola are drawn through a fixed point. Shew that the locus of their middle points is a hyperbola, similar to the original hyperbola or to its conjugate. [Joh. 1883.

83. On a plane field the crack of the rifle and the thud of the ball striking the target are heard at the same instant; find the locus of the hearer. [Joh. 1884.

84. In a rectangular hyperbola if PQ be a chord and CV the diameter conjugate to PQ, the angle between PQ and the tangent at P is equal to the angle VCP. [Sel. 1884.

85. From a point K on the conjugate hyperbola $KQPpq$ is drawn to meet the hyperbola in P, p and the asymptotes in Q, q: shew that $KP \cdot Kp = 2KQ \cdot Kq$. [Pet. 1883.

86. P, Q are two points on a hyperbola, through P is drawn a parallel to one asymptote and through Q a parallel to the other meeting the former parallel in T, the tangents at P and Q meet TQ, TP respectively in p, q; shew that pq is parallel to PQ. [Pet. 1883.

87. Let S, S' be the foci of a hyperbola, X, X' the points where the corresponding directrices meet SS', SY, $S'Y'$ the perpendiculars on a tangent, then if XY, $X'Y'$ meet the auxiliary circle again in y, y' shew that yy' is also a tangent to the hyperbola. [Pet. 1883.

88. If through each of the middle points of two chords of a rectangular hyperbola a parallel is drawn to the other, their intersection, the centre and the two middle points are on a circle. [Clare, 1883.

89. If through two vertices of a triangle inscribed in a hyperbola two lines be drawn parallel to the asymptotes to meet the opposite sides, the line which joins the points of

intersection will be parallel to the tangent at the third vertex. [CLARE, 1883.

90. If QV be an ordinate to the diameter PCp of a rectangular hyperbola, prove that QV is the tangent at Q to the circle round the triangle PQp. [T. H. 1883.

GENERAL CONICS.

1. S and H are the foci of a conic respectively corresponding to its two directrices, which latter are respectively intersected by a tangent to the conic in the points L and M. If N be the intersection of LS and MH (produced if necessary), prove that $LN = MN$. [I. C. S. 1885.

2. Given the focus and two points of a conic section, prove that the locus of the foot of the directrix is a circle.
[I. C. S. 1884.

3. In a central conic let PK, PL be the tangent and normal to the curve at P, and let KSL be drawn parallel to $S'P$, where S and S' are the foci. Prove that $KS = SL$.
[PET. 1887.

4. The tangent at P meets the major axis in T, perpendiculars to the axis from the feet of the perpendiculars through the foci to the tangent meet the curve in L, L' respectively: prove that TLL' are in a straight line.
[CLARE &c. 1888.

5. A straight line moves so that the intercept made on it by two fixed straight lines subtends a constant angle at a fixed point, shew that it touches a conic having this point as a focus. [TRIN. 1888.

6. If AB are two points of any diameter of a central conic section, and C, D two points on the conjugate diameter, prove that if the pole of AC lies on BD then also the pole of AD lies on BC. [LOND. 1st B.A., HON. 1870.

7. Prove that if two triangles are circumscribed about one conic they are inscribed in another.
[LOND. 1st B.A., HON. 1876.

C. G. 13

8. If any number of circles touch a conic at the same point; prove that the chords joining the points of intersection are all parallel. [LOND. 2nd B.A. 1873.

9. A series of conics have a common focus and directrix. Any straight line drawn at right angles to the directrix meets the conics in points P, Q, R.... Prove that the feet of the perpendiculars drawn from the common focus on the tangents at P, Q, R ... all lie on a straight line passing through the foot of the directrix. [JES. &c. 1888.

10. Shew that the locus of either extremity of the major axis of an ellipse inscribed in an isosceles triangle with that major axis parallel to the base, is a parabola with its vertex at the middle point of the perpendicular on the base from the vertex of the triangle. [JES. &c. 1888.

11. Two conics have a focus and directrix in common; and P, Q are two points, one on each conic, such that the angle PSQ is constant and equal to α. Prove that the tangents at P and Q intersect on a conic with the same focus and directrix. [JOH. 1887.

12. Prove that, if the lines joining to the foci any point P on a conic meet the conic again in Q and R, the line QR is always a tangent to a concentric and coaxial conic. [JOH. 1887.

13. The tangent at a moveable point P of a conic intersects a fixed tangent in Q, and from S the focus a straight line is drawn perpendicular to SQ and meeting in R the tangent at P; shew that the locus of R is a straight line. [JOH. 1888.

14. The tangent at any point P of a conic cuts the transverse axis in T and S is the focus; prove that the conic is an ellipse, a parabola, or a hyperbola, according as ST is greater than, equal to, or less than SP. [TRIN. 1886.

15. C is the centre of a given conic, O is a given point, and CO meets the conic in a point between C and O; a straight line $OPRQ$ meets the conic in P and Q, and the diameter conjugate to CO in a point R between P and Q; prove that $\dfrac{RP}{PO} - \dfrac{RQ}{QO}$ is independent of the direction of $OPRQ$. [TRIN. 1886.

16. Prove that the locus of the points of contact of parallel tangents to a series of confocal conics is a rectangular hyperbola passing through the foci of the confocals.
[TRIN. 1887.]

17. A conic has a given focus S, and a given focal chord PSQ. If the normal at P cuts the axis in G, find the locus of G.
[PEMB. 1886.

18. A conic is described passing through a given point P and having at that point a fixed tangent PT. The major axis is perpendicular to a fixed line PU and is equal to a given line. Shew that the centre lies on a hyperbola whose asymptotes are PU, PT.

19. If P be any point on a conic, PK the perpendicular on the directrix and KP be produced until PQ is equal to the focal distance of P, then the locus of Q is another conic.
[CATH. 1887.

20. Give a linear plane geometrical construction for drawing the common tangents of two conics which have at least two real points of intersection.
[JOH. 1886.

21. Spheres are drawn passing through a fixed point and touching two given planes. Prove that the points of contact lie on two circles, and that the locus of the centre of the sphere is an ellipse.

If the angle between the planes is the angle of an equilateral triangle, prove that the distance between the foci of the ellipse is half the major axis.
[JOH. 1887.

22. TP, PQ are two tangents to a conic, focus S, cutting the corresponding directrix in L, M respectively: prove that TS bisects the angle LSM.
[PET. 1885.

23. Given one of the foci of a conic inscribed in a triangle, shew how to find the other focus. Is more than one solution possible?
[PET. 1885.

24. Prove that the locus of the middle points of focal chords of a conic section is a similar conic section.
[PET. 1886.

25. Two similar and similarly situated conics intersect in A, B. A common tangent meets them in P, Q, and PQ is produced to a point R, so that $QR = PQ$. If RA,

RB meet the conic through P in H, K, and if HK meet QP produced in S, prove that $PS = PQ$. [Pet. 1886.]

26. A conic circumscribes a triangle ABC, and one focus lies on BC, find the envelope of the corresponding directrix. If A be a right angle shew that the envelope is a parabola whose focus is A and directrix BC. [Trin. 1885.]

27. Prove that if A, B and C are three given points, two parabolas can be drawn through A and B with C as focus, and that the axes of these parabolas are parallel to the asymptotes of the hyperbola which can be drawn through C with its foci at A and B. [Trin. 1886.]

28. If two conics have a common directrix their four points of intersection lie on a circle. [Caius, 1885.]

29. Prove that the locus of the intersection of tangents to an ellipse which make equal angles with the major and minor axes respectively, and are not at right angles is a rectangular hyperbola whose vertices are the foci of the ellipse. [Chr. &c. 1885.]

30. The asymptote CP of an hyperbola intersects an ellipse whose major and minor axes are respectively its conjugate and transverse axes in the point P: shew that if CP be produced to P' so that $PP' = CP$, and PM, $P'QM'$ be drawn perpendicular to CA meeting it in M, M' respectively, Q being the intersection of $P'QM'$ and the hyperbola, QM is the tangent at Q. [Sid. 1861.]

31. The two pairs of common tangents to two similar and similarly situated ellipses intersect in S, S', and are cut by a tangent to one ellipse in VT, VT' and by a tangent to the other in vt, $v't'$. Shew that if $V't'$ pass through S, $T'v'$ will also pass through S. [Trin. 1861.]

32. A parabola and a central conic intersect in four points, A, B, C, D; prove that the axis of the parabola is parallel to one of the lines joining the extremities of the diameters of the conic which are parallel to AB and CD. [Joh. 1861.]

33. The tangents at two points P, Q of a conic meet in O, and from O are drawn two straight lines cutting the

conic and making equal angles with the transverse axis. If they meet PQ in M, N, and the middle points of the chords be R, S, shew that $RMNS$ lie on a circle.

[PET. 1882.]

34. Two conics have their directrices parallel, and the same focus S: if any straight line through S meet the two conics in P and Q, find the locus of the middle point of PQ.

[CHR. 1882.]

35. A, B, C are any three fixed points; through A any straight line is drawn which cuts a given conic in the points P, Q. Shew that the locus of the intersection of PB and QC is a conic.

[JES. 1886.]

36. O is a fixed point, and P any point on a given straight line. PQ is taken along the line always in a constant ratio to OP. Prove that the line joining P to the middle point of OQ always touches a conic whose focus is O.

[JES. 1886.]

37. Prove that if an ellipse and a hyperbola are confocal they intersect each other at right angles, and that the asymptotes of the hyperbola pass through the points on the auxiliary circle of the ellipse which correspond to the points of intersection.

[JOH. 1886.]

38. A line AB is drawn from a fixed point A to meet a fixed circle in B: through B a line BC is drawn perpendicular to AB, to meet a concentric circle in C. Shew that a line through C parallel to AB touches a conic.

[PET. 1884.]

39. Two tangents are drawn from a point on the directrix to a central conic, and the points of contact joined. Shew that the locus of the orthocentre of the triangle thus formed is a conic similar to the given one.

[PET. 1884.]

40. A fixed straight line meets one of a system of confocal conics in two points. Prove that the locus of the point where the normals at these points intersect is a straight line.

[PET. 1884.]

41. With any point on the directrix of a given parabola as focus and the focus of the parabola as the other focus, an ellipse or hyperbola is described, shew that the tangents and normals at its points of intersection with the directrix are also tangents to the parabola.

[PET. 1884.]

42. A fixed chord PQ of a conic meets any diameter in N, and the ordinate to this diameter through N meets the tangents at P and Q in H, K. Prove that HK is bisected at N. [Caius, 1883.

43. If any two chords PQ, PQ' be drawn through a point P of a conic and perpendiculars to the chord through Q and Q' meet the normal at P in N, N' respectively, shew that PN, PN' are to one another as the squares of the diameters of the conic parallel to PQ, PQ'. [Pet. 1885.

44. If A, B, C, D are four points on a conic the normals at which meet in a point, prove that the sum of the squares of the diameters parallel to AB and CD is equal to the sum of the squares of the diameters parallel to AC and BD.
[Clare, 1885.

45. A parabola passes through two fixed points A, B at a distance $2a$ apart, and has a straight line distant c from the middle point of AB as directrix. Shew that the locus of the focus of the parabola is a conic section, which is an ellipse or a hyperbola, according as c is greater or less than a. [Trin. 1884.

46. A circle is drawn on a sheet of paper and the paper is folded so that one corner of the sheet lies on the circumference of the circle. Prove that as this corner moves about on the circle the crease on the paper will envelope a conic. [Trin. 1884.

47. A semicircular piece of paper is folded over so that a particular point P on the bounding diameter lies on the circular boundary; prove that the crease-line touches a fixed conic. [Trin. 1885.

48. If a circle and a conic intersect in the points B, C, D, E then the lines bisecting the angles between BC and DE, BD and CE, BE and CD are each parallel to one of two given straight lines. [Caius, 1885.

49. TP, TP' are tangents to a conic, PG, $P'G'$ are normals at P, P': prove that $TP : TP' :: PG : P'G'$. Prove also that if GL, $G'L'$ are drawn perpendicular to PP', then
$$PL = P'L'.$$ [Chr. 1885.

PROBLEMS.

50. Two tangents to a conic are drawn from any point T touching the conic in P and Q, any straight line drawn parallel to TP meets TQ in L, PQ in M and the conic in R, S: shew that $LO^2 = LR \cdot LS$. [Qu. 1885.

51. P, Q are any two points on an ellipse whose foci are S, H; SP, HQ intersect in M, SQ, HP in N, and the bisectors of the angles QSP, QHP in R. Shew that RP, RQ are tangents to the ellipse, and M, N are points on a confocal hyperbola to which RM, RN are tangents.
[Jes. 1885.

52. Given a line, a circle with centre O, and a point S: a variable point R on the line is joined to S by a line which meets the circle in U, V, and lines are drawn from S parallel to OU, OV to meet RO in points P and Q; shew that the locus of these points is a conic with S as focus and the given line as directrix.

Deduce from this mode of generation that tangents from any point to a conic subtend equal angles at a focus.
[Joh. 1884.

53. Prove that the diagonals of a curvilinear quadrilateral formed by the intersection of two confocal ellipses with two confocal hyperbolas are equal.

Shew that these results are also true for a system of confocal and coaxial parabolas. [Joh. 1884.

54. A hyperbola is described having a focus of an ellipse for focus, and the tangent at the corresponding vertex for directrix. Prove that tangents to the ellipse from points in which the hyperbola cuts the minor axis of the ellipse are parallel to the asymptotes of the hyperbola.
[Joh. 1884.

55. An ellipse and a hyperbola have the same foci and meet in P. PYZ is a tangent to the hyperbola at P; $SY \cdot HZ$ the focal perpendiculars. Prove that

$$PY \cdot PZ = BC^2,$$

where BCB' is the minor axis of the ellipse. [Pet. 1884.

56. An ellipse is met in P and Q by a rectangular hyperbola having for asymptotes the axes of the ellipse.

PM, QN are ordinates drawn to the axis CA; PR, QT to CB. Prove that
$$CM^2 + CN^2 = CA^2,$$
and that $\quad CN : CR :: CA : CB$. [Pet. 1884.

57. From a fixed point O on the circumference of a circle a chord OA is drawn, and produced to B so that the difference of the squares on OB and OA is constant, prove that the line through B perpendicular to OB will touch a conic of which O is centre and the other extremity of the diameter of the circle through O is a focus. [Clare, 1884.

58. Given a focus S and two tangents to a conic, prove that the envelope of the minor axis is a parabola of which the focus is S. [Trin. 1884.

59. A focal chord PSQ of a conic is given in position and the position of the axis is also given. Trace the conic.
[Pemb. 1884.

60. Prove by projection that, if ACA' be the major axis of an ellipse, and PNP' a double ordinate bisecting CA' at N, the tangent at P is parallel to AP'.
[Pemb. 1884.

61. An ellipse and a hyperbola are concentric and co-axial, and a point P is such that its polars with respect to the two are at right angles and intersect in Q; prove that the locus of P is two straight lines through the centre C, and the locus of Q is two other straight lines through the centre; but that if the conics be confocal, C, Q and P are in one straight line and $CP \cdot CQ$ is constant. [Chr. 1884.

62. Given the focus, directrix and eccentricity, give a geometrical construction for the points where a given straight line drawn through the focus cuts the curve.
[Qu. 1884.

63. PQ is any chord of a conic, PG, QH the normals, G, H being on the axis, GL, HK are perpendiculars on PQ, shew that $PL = QK$. [Cath. 1884.

64. Prove that if A, B, C are three given points, two parabolas can be drawn through A and B with C as focus, and that the axes of these parabolas are parallel to the asymptotes of the hyperbola which can be drawn through C with its foci at A and B. [Trin. 1885.

65. If a parabola, having its focus coincident with one of the foci of an ellipse, touches the conjugate axis of the ellipse, a common tangent to the ellipse and parabola will subtend a right angle at the focus. [Trin. 1885.

66. ACA' and BCB' are the transverse and conjugate axes of an ellipse, of which S and S' are the foci, P is one of the points of intersection of this ellipse and a confocal hyperbola, and aCa' is the transverse axis of the hyperbola. Prove that $SP = Aa$, $S'P = A'a$, and $aB = CP$.
[Trin. 1885.

67. Two fixed points P, Q are taken in the plane of a given circle, and a chord RS of the circle is drawn parallel to PQ, prove that for different positions of RS the locus of the point of intersection of RP and SQ is a conic. [Trin. 1886.

68. A circle passes through a fixed point and cuts a given straight line at a constant angle. Prove that the locus of the centre is a conic. [Jes. 1884.

69. A chord of a conic subtends a given angle at the focus. Prove that the tangents at its extremities will intersect on a conic having the same focus and directrix as the original conic. [Joh. 1883.

70. An ellipse and hyperbola have the same transverse axis, and their eccentricities are the reciprocals of one another; prove that the tangents to each through the focus of the other intersect at right angles in two points and also meet the conjugate axes on the auxiliary circle.
[Joh. 1884.

71. From any point Q on a central conic, QS, QH are drawn to the foci S, H, meeting the conic again in P, P'; show that if the tangents at P, P' meet in T, QT is bisected by the minor axis and the locus of T is a conic.
[Pet. 1883.

72. Through two points on a central conic shew that two circles can be described to touch the conic; and that the points of contact are at the extremities of a diameter.
[Caius, 1883.

CONE.

1. If S be a point within the cone; A its vertex, AB its axis; shew that the difference of the acute angles made with AB by the planes of the sections having S for a focus is twice the angle SAB. [I. C. S. 1887.

2. Shew how to obtain from a given cone a section which shall have the greatest possible eccentricity.
[I. C. S. 1886.

3. Under what circumstances may the section of a cone by a plane be a rectangular hyperbola? In such a case shew how to determine the necessary inclination of the cutting plane. [I. C. S. 1885.

4. Shew how to find the centre and the asymptotes of a hyperbolic section of a cone. Also shew how to cut from a given cone a hyperbola, whose asymptotes shall contain the greatest possible angle. [I. C. S. 1884.

5. Prove that the minor axis of an elliptic section of a right cone is a mean proportional between the diameters of the circular sections of the cone, made by planes drawn through the extremities of the major axis of the ellipse.

If the ellipse be projected upon a plane perpendicular to the axis of the cone, shew that the distance between the foci of the curve of projection is equal to the difference between the radii of the same two circular sections.

6. From a given right circular cone is cut a series of parabolas the axes of which intersect a given straight line OM which passes through the vertex O. If any section intersect OM at N, shew that the ratio $ON^2 : AN.CL$ is constant for all the parabolas, where A is the vertex of the section and C the centre of its focal sphere, and L is the point where the section cuts the axis OL of the cone.
[Pemb. 1887.

7. If two sections of a cone have a common directrix, the latera recta of the sections are in the ratio of their eccentricities. [Jes. &c. 1888.

8. Prove that the locus of the centres of all plane sections, for which the distance between the foci is the same, is a right circular cylinder. [Joh. 1888.

9. Prove that the centres of all sections having their minor axis of the same length lie on the surface formed by a hyperbola revolving about its transverse axis. [Pet. 1887.

10. What conditions are necessary in order that it may be possible to construct an elliptical cone passing through two given circles in different planes? [Trin. 1887.

11. Show that the locus of the vertices of all right cones out of which an ellipse given both in magnitude and position can be cut, is a hyperbola passing through the foci of the ellipse. [Jes. 1887.

12. Show how to draw a plane cutting a given right cone in an ellipse of given eccentricity and having a major axis of given length. [Cath. 1887.

13. If the vertical angle of a cone be a right angle, show that the square of the sum of the radii of the two contact spheres of a section by a plane is equal to the sum of the squares of the axes of the section. [Pet. 1886.

14. Two right circular cones whose vertical angles are right angles, have their vertices and one generating line coincident, prove that when a section of each is made by the same plane, the minor axis of the one section is equal to the conjugate axis of the other. [Clare, 1886.

15. Prove that the latera recta of parabolic sections of a right circular cone are proportional to the distances of their vertices from the vertex of the cone. [Trin. 1886.

16. Through a fixed rectangular hyperbola a series of right circular cones is described. Prove that the locus of their vertices is an ellipse with eccentricity $\dfrac{1}{\sqrt{2}}$.
[Pemb. 1885.

17. If P be a common point of two intersecting spheres which are inscribed in a right cone, show that the tangent planes at P will make equal angles with the straight line drawn from P to the vertex of the cone. [T. H. 1886.

18. Any section of a right circular cylinder by a plane not parallel or perpendicular to its axis is an ellipse.
[Qu. 1886.

19. Different elliptic sections of a right cone are taken such that their axes are equal (the major axes all being in one plane). Show that the locus of their centres is a hyperbola. [Cath. 1886.

20. Determine the parabolic section of a given cone, which shall have its latus-rectum of a given magnitude.
[T. H. 1881.

21. Prove that the semi minor axis of an elliptic section of a right cone is a mean proportional between the perpendiculars drawn from the vertices of the ellipse upon the axis of the cone. If V be the vertex of the cone, R the point where the axis of the cone cuts AA', the major axis of the section, prove that
$$CR : CA :: CS : AV + CS.$$ [Trin. 1861.

22. A series of elliptic sections of a right circular cone are made by parallel planes; shew that the auxiliary circles lie on a right cone having for its base an ellipse similar to the given ellipses. [T. H. 1882.

23. Two cones have their vertical angles supplementary; prove that the sum of the squares of the reciprocals of the greatest eccentricities of conics, obtained from them by plane sections, is unity. [Trin. 1885.

24. Shew how to draw a section which shall have a given straight line for directrix, the given straight line being perpendicular to the axis of the cone. [Qu. 1885.

25. Given an ellipse and a right circular cone, place the ellipse so as to be a plane section of the cone. [Trin. 1884.

26. Prove that the latus-rectum of a plane section of a cone varies as the perpendicular from the vertex of the cone upon the plane of section. [Trin. 1884.

27. If two different plane sections of a cone have a common directrix the line joining their foci goes through the vertex of the cone. [Qu. 1884.

28. If the angle of a cone be a right angle, prove that the semi-latus-rectum of a section is a mean proportional between the segments of the major axis made by a perpendicular on it from the vertex of the cone. [Cath. 1884.

29. Two cones which have a common vertex, their axes at right angles, and their vertical angles supplementary are intersected by a plane at right angles to the plane of their axes. Prove that the distances of either focus of the elliptic section from the foci of the hyperbolic section are equal respectively to the distance from the vertex of the ends of the transverse axis of each, and that the sum of the squares on the semi-conjugate axes is equal to the rectangle contained by these distances. [TRIN. 1885.

30. If the minor axis of the section of a cone be constant, prove that the centre of it lies on a hyperboloid of revolution. [JES. 1884.

APPENDIX.

ELLIPSE.

Proposition I. (continued).

To prove that *the curve lies between lines drawn through* A *and* A' *at right angles to the axis.*

On SN or SN produced mark off $SK = e \cdot XN$.

We must consider in what positions of N, NP meets the circle whose centre is S and radius $e \cdot XN$; i.e. whether SK is greater or less than SN.

Case 1. If N is between S and A.

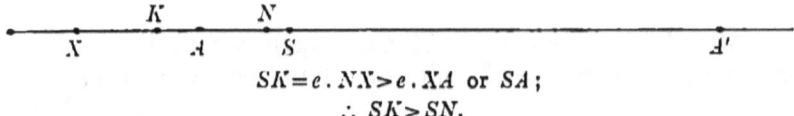

$$SK = e \cdot NX > e \cdot XA \text{ or } SA;$$
$$\therefore SK > SN.$$

Case 2. If N is between S and A'.

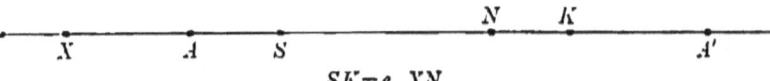

$$SK = e \cdot XN,$$
and $$SA' = e \cdot XA';$$
$$\therefore \text{ by subtraction } KA' = e \cdot NA' < NA';$$
$$\therefore SK > SN.$$

Case 3. If N is in SA' produced.

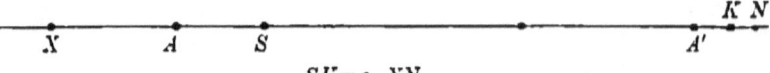

$$SK = e \cdot XN,$$
and $$SA' = e \cdot XA';$$
$$\therefore \text{ by subtraction } A'K = e \cdot A'N < A'N;$$
$$\therefore SK < SN.$$

Case 4. If N is between A and X.

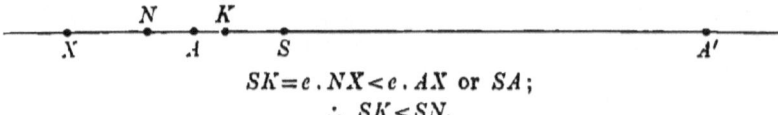

$$SK = e \cdot NX < e \cdot AX \text{ or } SA;$$
$$\therefore SK < SN.$$

Case 5. If N is in SX produced.

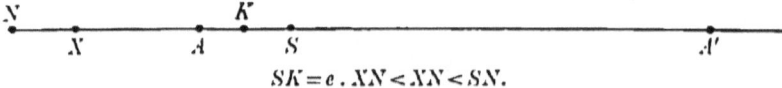

$$SK = e \cdot XN < XN < SN.$$

We have now proved that the circle intersects the perpendicular NP, when N is in any part of the axis AA' between A and A', but not when N lies outside the part AA', hence the ellipse lies entirely between lines drawn through A and A' at right angles to the axis.

HYPERBOLA.

PROPOSITION I. (*continued*).

To prove that *the curve lies outside lines drawn through* A *and* A' *at right angles to the axis.*

On SN or SN produced mark off $SK = e \cdot XN$.

We must consider in what positions of N, NP meets the circle whose centre is S and radius $e \cdot NX$; i.e. whether SK is greater or less than SN.

Case 1. If N is between A and X.

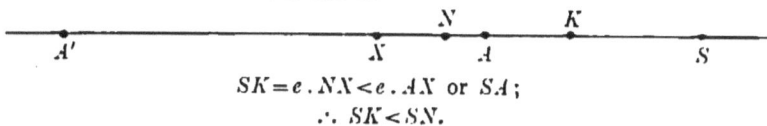

$$SK = e \cdot NX < e \cdot AX \text{ or } SA;$$
$$\therefore SK < SN.$$

Case 2. If N is between X and A'.

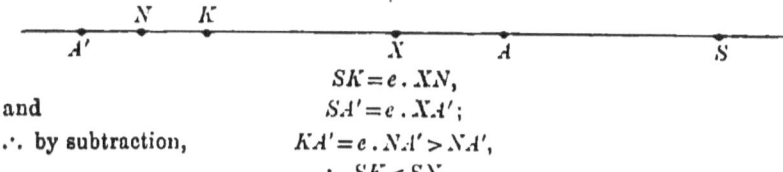

$$SK = e \cdot XN,$$
and $$SA' = e \cdot XA';$$
$\therefore$ by subtraction, $$KA' = e \cdot NA' > NA',$$
$$\therefore SK < SN.$$

Case 3. If N is in SA' produced.

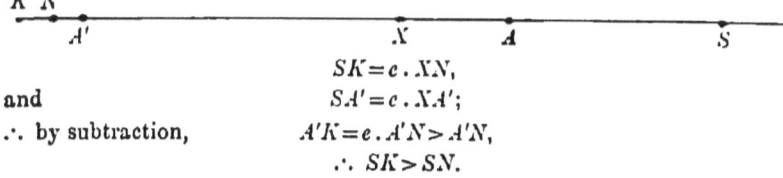

$$SK = e \cdot XN,$$
and $$SA' = e \cdot XA';$$
$\therefore$ by subtraction, $$A'K = e \cdot A'N > A'N,$$
$$\therefore SK > SN.$$

Case 4. If N is between A and S.

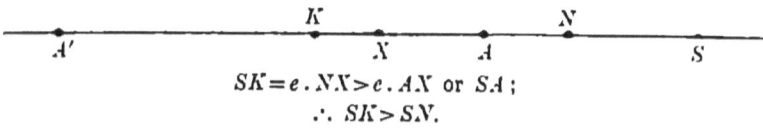

$$SK = e \cdot NX > e \cdot AX \text{ or } SA;$$
$$\therefore SK > SN.$$

Case 5. If N is in AS produced.

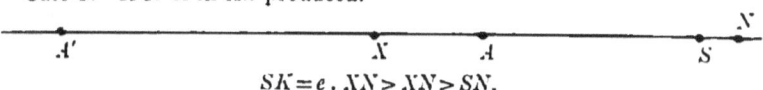

$$SK = e \cdot XN > XN > SN.$$

We have now proved the circle does not intersect the perpendicular NP, when N is in any part the axis AA' between A and A', but they do intersect when N lies outside the part AA', hence the hyperbola lies entirely outside the lines drawn through A and A' at right angles to the axis.

Cambridge:
PRINTED BY C. J. CLAY, M.A. AND SONS,
AT THE UNIVERSITY PRESS.

www.ingramcontent.com/pod-product-compliance
Lightning Source LLC
Chambersburg PA
CBHW020822230426
43666CB00007B/1065